Hybrid Circuit Design and Manufacture

ELECTRICAL ENGINEERING AND ELECTRONICS

A Series of Reference Books and Textbooks

Editors

Marlin O. Thurston
Department of Electrical
Engineering
The Ohio State University
Columbus, Ohio

William Middendorf
Department of Electrical
and Computer Engineering
University of Cincinnati
Cincinnati, Ohio

1. Rational Fault Analysis, *edited by Richard Saeks and S. R. Liberty*
2. Nonparametric Methods in Communications, *edited by P. Papantoni-Kazakos and Dimitri Kazakos*
3. Interactive Pattern Recognition, *Yi-tzuu Chien*
4. Solid-State Electronics, *Lawrence E. Murr*
5. Electronic, Magnetic, and Thermal Properties of Solid Materials, *Klaus Schröder*
6. Magnetic-Bubble Memory Technology, *Hsu Chang*
7. Transformer and Inductor Design Handbook, *Colonel Wm. T. McLyman*
8. Electromagnetics: Classical and Modern Theory and Applications, *Samuel Seely and Alexander D. Poularikas*
9. One-Dimensional Digital Signal Processing, *Chi-Tsong Chen*
10. Interconnected Dynamical Systems, *Raymond A. DeCarlo and Richard Saeks*
11. Modern Digital Control Systems, *Raymond G. Jacquot*
12. Hybrid Circuit Design and Manufacture, *Roydn D. Jones*

Other Volumes in Preparation

Hybrid Circuit Design and Manufacture

ROYDN D. JONES
Tektronix Laboratories
Beaverton, Oregon

Sponsored by International Society
for Hybrid Microelectronics

MARCEL DEKKER, INC.　　　　　New York and Basel

To my wife and parents

Library of Congress Cataloging in Publication Data

Jones, Roydn D., [date]
  Hybrid circuit design and manufacture.

  (Electrical engineering and electronics ; 13)
  Includes index.
  1. Hybrid integrated circuits. 2. Thin-film circuits.
3. Thick-film circuits. 4. Electronic circuit design.
I. Title. II. Series.
TK7874.J66   621.381'73   81-17535
ISBN 0-8247-1689-2   AACR2

COPYRIGHT © 1982 by MARCEL DEKKER, INC. ALL RIGHTS RESERVED

Neither this book nor any part may be reproduced or transmitted in any form or by any means, electronic or mechanical, including photocopying, microfilming, and recording, or by any information storage and retrieval system, without permission in writing from the publisher.

MARCEL DEKKER, INC.
270 Madison Avenue, New York, New York 10016

Current printing (last digit):
10 9 8 7 6 5

PRINTED IN THE UNITED STATES OF AMERICA

FOREWORD

The International Society For Hybrid Microelectronics (ISHM) is pleased to sponsor this volume, *Hybrid Circuit Design and Manufacture*. Since it was founded in 1967, this society has held as one of its major goals the interchange of technical information. In past years this has been accomplished mainly through local chapter meetings, the annual Symposium, and the technical Journal. As the diversity of applications and technology of hybrid microcircuits has expanded, the need for tutorial and guideline textbooks has become more urgent: for, although individual technical papers published in the Journal and Symposium Proceedings are current at the time of publication, there needs to be a periodic condensation and culling of obsolete information, which best can be presented in textbook form.

    That hybrid technology has reached a level of maturity so as to warrant textbooks no doubt comes as a surprise to those who in the recent past have predicted the demise of this field. The hybrid specialty was created in response to semiconductor chip integration and packaging needs. Why then has it persevered in a period during which IC technology is becoming more integrated in monolithic form, with larger and more complex semiconductor chips increasingly assuming the functions of individual hybrids? The answer, of course, lies in the continuing need to integrate these more complex IC chips into the next level of system assembly. Hybrid technology has, in fact, grown well beyond the initial need for interconnection of passive and active devices (the definition of "hybrid"), and has assumed the stature of a specialty within the field of microelectronic packaging. Two characteristics of this identity are notable: hybrid microcircuits are becoming increasingly complex, and their usage actually is encroaching on next-level packaging technologies, e.g., printed wiring boards.

    Basic tools of the hybrid trade comprise the materials and processes of microelectronic device interconnection and packaging. In this book, Roydn Jones gives very detailed descriptions of these tools, thus providing a single text of many uses. Although principally directed toward hybrid microcircuit designers, it is suitably composed to act as an engineering textbook. It will be useful as a management primer as well, since it also includes information on reliability and production.

ISHM is grateful to the author and publisher for this opportunity to encourage distribution of technical information in the dynamic field of hybrid microelectronics.

Richard P. Himmel
Technical Vice President
International Society For Hybrid Microelectronics

PREFACE

Hybrid technology covers a wide range of disciplines, including material science, ceramics technology, physics, chemistry, and electronic engineering. This book is aimed primarily at electronic engineering designers and design managers who seek a background in hybrid technology. The book differs from many in the hybrid field as it covers both thick- and thin-film hybrids.

The introductory chapter discusses the electrical properties and economic aspects of hybrids compared with discrete and integrated circuit technologies. Chapter 2 deals with substrate materials used in both thick and thin films. Chapters 3 through 6 cover thick-film and thin-film materials and design guidelines.

Laser trimming is a key to precision hybrid circuitry and is covered in Chapter 7, including design curves for typical trim geometries.

Hybrid assembly techniques are covered in Chapter 8, including an overview of available hybrid devices, soldering techniques, chip and wire attachment, and beam lead and tape automated bonding.

Production and quality assurance are dealt with in Chapter 9, and thermal design considerations are explained from a theoretical point of view in Chapter 10.

Chapters 11 and 12 include discussions of hybrid circuit partitioning and the overall hybrid design cycle. The typical performance characteristics and applications of thick and thin films are compared.

The final chapter examines the reliability aspects of hybrid circuits, obviously a very important aspect of any technology.

I would like to acknowledge the numerous colleagues who have contributed in some way to this work. In particular I am indebted to Lance Mills, who first introduced me to thin-film hybrids at Hewlett-Packard, and to Bob Coackley, who subsequently believed in me enough to let me develop my knowledge of hybrids at Hewlett-Packard;

also Bob Holmes at Tektronix for his encouragement over the years, and to Art Seidman, who originally suggested this work. Finally, I am very grateful to my wife, Ursel, for her help and support in this venture, particularly for typing of the manuscript.

Roydn D. Jones

CONTENTS

Foreword, Richard P. Himmel   *iii*

Preface   *v*

1. Introduction to Hybrids   *1*

    1.1 Introduction   *1*
    1.2 What Is a Hybrid?   *1*
    1.3 Why Hybrids?   *1*
    1.4 Where Are Hybrids Used?   *4*
    1.5 Comparison with Discrete and Integrated Circuit Technologies   *4*
    1.6 Hybrid Film Technologies   *6*

2. Substrate Materials and Properties   *9*

    2.1 Substrate Properties   *9*
    2.2 Substrate Materials   *10*
    2.3 Electrical Substrate Specifications   *13*
    2.4 Mechanical Substrate Specifications   *17*
    2.5 Substrate Fabrication: Alumina   *19*

3. Thick-Film Materials   *22*

    3.1 Introduction   *22*
    3.2 Thick-Film Conductors   *22*
    3.3 Thick-Film Dielectrics   *27*
    3.4 Thick-Film Resistors   *29*
    3.5 Thick-Film Processing   *31*

4. Thick-Film Design Guidelines   37

   4.1 General Design Procedures   37
   4.2 Screening Reference Corner and Screening Procedure   37
   4.3 Basic Design Rules for Thick-Film Resistors   40
   4.4 Basic Design Rules for Thick-Film Conductors   48
   4.5 Basic Design Rules for Thick-Film Capacitors   58

5. Thin-Film Materials   62

   5.1 Introduction   62
   5.2 Thin-Film Processes   62
   5.3 Thin-Film Conductor Materials   66
   5.4 Thin-Film Resistor Materials   67
   5.5 Thin-Film Dielectric Materials   67
   5.6 Thin-Film Processing (Conductor-Resistor Networks)   67
   5.7 Environmental Protection of Thin-Film Resistors   71

6. Thin-Film Design Guidelines   72

   6.1 Description of Thin-Film Resistors   72
   6.2 Thin-Film Conductor Guidelines   75
   6.3 Description of Thin-Film Capacitors and Crossovers   75
   6.4 Thin-Film Snapstrates   79

7. Laser Trimming of Hybrid Circuits   81

   7.1 The Need for Trimming   81
   7.2 The Laser Trimming Process   82
   7.3 Laser Trim System   83
   7.4 Resistance Measurement Techniques   85
   7.5 Trim Information for Top-Hat Resistors   88

8. Assembly Techniques   95

   8.1 Introduction   95
   8.2 Components Used in Hybrid Assembly   95
   8.3 Assembly Techniques   103
   8.4 Packaging   115

9. Production Considerations   118

   9.1 Documentation   118
   9.2 Inspection and Reliability   119
   9.3 Technical Support   120
   9.4 Labor Aspects   120
   9.5 Yields   122
   9.6 In-House or External Manufacturing   122

Contents

10. Thermal Considerations *124*

    10.1 Introduction *124*
    10.2 Cooling Phenomena *124*
    10.3 Cooling of Hybrid Microcircuits *125*
    10.4 Analytical Techniques *126*
    10.5 Thermal Conductivity of Materials *133*
    10.6 Film Resistor Power Densities *134*
    10.7 Factors for Good Thermal Design *136*

11. Circuit Partitioning *138*

    11.1 Partitioning *138*
    11.2 Monolithic, Hybrid, or Discrete? *139*
    11.3 Thick or Thin Film? *141*

12. Design Cycle *144*

    12.1 Design Cycle *144*
    12.2 Circuit Analysis and Simulation *144*
    12.3 Layout *146*
    12.4 Processing and Packaging *147*
    12.5 Test and Redesign *147*
    12.6 Reliability and Life Tests *148*
    12.7 Testing of Hybrid Circuits *149*
    12.8 Overall Design Cycle *150*

13. Reliability *152*

    13.1 Reliability Aspects of Hybrids *152*
    13.2 Reliability Data *152*
    13.3 Reliability Definitions *153*
    13.4 Constant Failure Rate *155*
    13.5 Failure Rate of Multiple-Element Systems *158*
    13.6 Establishing Reliability Goals *158*
    13.7 Reliability Testing *160*
    13.8 Failure Mechanisms in Hybrids *162*
    13.9 Failure Analysis *163*
    13.10 Other Reliability Tests *163*

Glossary of Terms 165

Index 209

1
INTRODUCTION TO HYBRIDS

1.1 INTRODUCTION

Webster's New Collegiate Dictionary defines a *hybrid* as "something heterogeneous in origin or composition," i.e., consisting of dissimilar ingredients or constituents. Due to this heterogeneity of hybrid microcircuits, and the history of their development, there are naturally numerous design approaches. The intention of this book is to provide a basic understanding of the design guidelines for a wide range of hybrid circuits, both thick and thin film, covering a wide range of frequencies.

1.2 WHAT IS A HYBRID?

A hybrid microcircuit is fabricated on an insulating substrate utilizing some combination of thick- or thin-film components, monolithic semiconductor devices, and discrete parts.

The substrate usually comprises an alumina or similar ceramic wafer on which is deposited or screened single or multiple layers of conductors, resistors, and capacitors. Discrete, but usually unencapsulated semiconductors, capacitors, resistors, and inductors are mechanically attached to the substrate. With discrete devices the attaching method usually provides the required electrical connection to the circuit. With semiconductor devices wire bonds are used to connect from device contact "pads" to the substrate. Various packaging and interconnect schemes can be used to connect the hybrid circuit to the external world. Figure 1.1 shows a variety of typical hybrid circuits.

1.3 WHY HYBRIDS?

In many applications hybrid circuitry is used because there is no alternative technology. This is particularly true in higher-frequency circuits and microwave circuits, where the small size, line definition,

FIGURE 1.1. Typical hybrid circuits.

Introduction to Hybrids

and reproducibility of hybrids come into their own. Hybrid technology has made major contributions to the wide range of low-cost, high-performance microwave components available today. Components in this category include wide-bandwidth amplifiers, microwave oscillators, filters, directional couplers, power splitters, and mixers. Such hybrid microwave integrated circuits may be either passive or active networks.

At lower frequencies hybrid resistor networks can be used cost effectively compared with their discrete counterparts. Such networks are finding widespread use in analog-to-digital and digital-to-analog converters. More complex active hybrids can also be justified at lower frequencies when precision, automatic adjustment, size, weight, and thermal design may be major considerations.

With constant advances in multilayer hybrids, packaging techniques, and diagnostic abilities it is becoming feasible to put many integrated circuits down on a single hybrid. This is finding applications in many areas, particularly in the design of minisystems and hand-held instruments, where size and weight are important.

Some of the advantages of using hybrids compared with discrete circuits are as follows:

Electrical properties
1. Higher-frequency performance
2. Higher density
3. Predictability of design
4. Long-term stability and reliability
5. Low-temperature coefficient of resistance
6. Small absolute and relative tolerances
7. Ability to trim components for both passive and functional response
8. High thermal conductivity of substrates

Economic Aspects
1. Miniaturization, leading to reduced weight and size
2. Repeatability and lower test times
3. Improved specifications on circuits
4. Lower cost for matched, close-tolerance resistors
5. Higher reliability with lower warranty costs
6. Easy serviceability and replaceability in the field
7. Relatively simple processing and assembly techniques
8. Low development cost

In any particular application many of these reasons may justify a hybrid circuit. Careful comparison among discrete, hybrid, and integrated circuit approaches is often required.

1.4 WHERE ARE HYBRIDS USED?

Hybrids are to be found in almost every area of consumer, industrial, and military electronics. Table 1.1 indicates areas where hybrids are particularly applicable.

It has been clearly shown that hybrids have found a home in high-frequency and microwave electronics. They are also being used more frequently to provide mother board interconnect systems for medium-scale and large-scale integrated circuits (MSI and LSI).

1.5 COMPARISON WITH DISCRETE AND INTEGRATED CIRCUIT TECHNOLOGIES

Printed or Etched Circuit Board Techniques

A *printed circuit board* (PCB) or *etched circuit board* (ECB) consists of a thin copper sheet laminated onto some kind of resin-bonded material. The copper is usually etched chemically to provide the interconnections for discrete components. Wave soldering is usually used to solder these components to the board.

The need for high package density on such boards has been partially achieved by the use of double-sided and multilayer boards. However, to achieve very high density, two basic problems still exist with such boards. One is that the components themselves are large, particularly with respect to the active part of the device. For example, a typical 0.6 × 0.3 in. dual-in-line package used on an ECB contains a silicon device usually smaller than 0.1 × 0.1 in. Second, the ECB itself is a poor thermal conductor, and dense circuitry usually entails dissipating large amounts of heat generated by active devices and resistors. This can be done on an ECB only by the use of complex heat sinking of devices, which adds cost and size to the design. Hybrids lead to much higher density packaging since semiconductor dice are bonded directly to the substrate, and resistors and capacitors are inherent in the films themselves. Typical hybrid substrates are significantly better thermal conductors than etched circuit boards, and bonding semiconductor dice directly to the substrate leads to excellent thermal designs.

Due to the size of components used on ECBs the lengths of component interconnects can also be a problem at high frequencies. Added inductance and capacitance can have a detrimental effect on frequency and phase characteristics in analog circuitry, and add to propagation delays and reflection problems in digital circuitry. The small size of hybrid circuitry and closer proximity of components usually leads to much improved designs over the equivalent ECB approach.

A major advantage of the ECB approach over hybrids is the ease of replacement of discrete components on the ECB. Integrated circuits,

TABLE 1.1. Applications Areas for Hybrid Microcircuits

Consumer	Industrial	Military
Television	Electronic instruments	Portable equipment
Radio	Electronic calculators	Communications equipment
Time pieces	Computers	Microwave equipment
Automotive products	Computer peripherals	Radar systems
Appliances	Medical instruments	Guidance systems
Electronic organs	Communications equipment	Navigation systems
Toys	Microwave equipment	Infrared detectors
	Controllers	Displays
	Digital equipment	Sonar systems
	Displays	Missiles
	A/D converters	Proximity fuses
	D/A converters	

resistors, etc., can be easily desoldered from the board and replaced, even in the field environment. With hybrids, however, component attachment involves more complex processes, such as wire bonding and die attach, and such facilities usually do not exist in the field. Thus, with a hybrid failure, the hybrid module itself needs to be replaced and the failed component scrapped or returned to the factory for repair. For this reason, reliability and replacement cost need to be major considerations in hybrid design.

Semiconductor Integrated Circuits

An *integrated circuit* (often called an IC) consists of transistors, diodes, resistors, and capacitors, usually diffused into a semiconductor material, usually silicon. Since all the components are fabricated on the same piece of silicon, integrated circuit construction is also called *monolithic*. The unpackaged silicon integrated circuit is often referred to as a chip or die. Almost any circuit, analog or digital, can be produced using IC technology, although it does have certain problems which can be overcome by using hybrids.

Certain components, such as large capacitors and inductors, are difficult to incorporate on an IC. However, these can be fabricated as films or added as discrete components on hybrids.

Also, as the size of ICs increase, two problems are often encountered. The yield drops off significantly as size increases beyond certain limits (although these limits are increasing rapidly as process technology improves). Thermal dissipation problems frequently occur as the device size increases, and often an optimum solution is to partition the circuit into several high-yielding, low-power ICs and interconnect them on a hybrid.

A further drawback of IC technology is the high development time and development cost of a typical IC. For low-volume custom designs it may be faster and more economical to design with off-the-shelf components on a hybrid than to commit to an IC development program. The hybrid development time and cost are usually significantly less than the equivalent for IC development.

1.6 HYBRID FILM TECHNOLOGIES

Two film technologies are encountered in the hybrid field: thick film and thin film. These technologies are introduced briefly here to differentiate between them, and are discussed in more detail in later chapters.

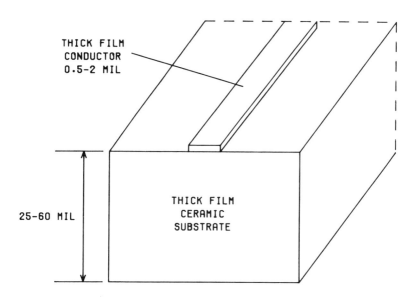

FIGURE 1.2. Thick-film dimensions.

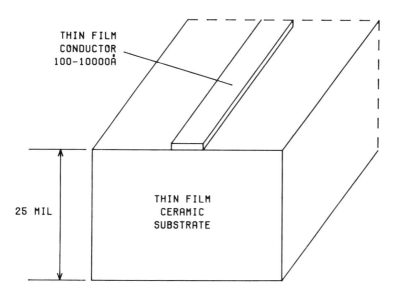

FIGURE 1.3. Thin-film dimensions.

Definition of Thick Film

Thick film is the preferred generic description for the field of micro-electronics in which specially formulated pastes are applied and fired onto a ceramic or insulating substrate in a definite pattern and sequence to produce a set of passive components. Conductors, resistors, capacitors, and inductors can be fabricated with thick film. The pastes are usually applied using a silk screen method. The high-temperature firing oxidizes or reduces the metallic elements to develop the required component characteristics.

The term thick film is derived from the fact that the fired films are fairly thick, being typically 10 to 50 μm (approximately 0.5 to 2 mil) in thickness. Figure 1.2 illustrates typical thick-film dimensions.

Definition of Thin Film

Thin film is the preferred generic description for the field of micro-electronics in which conductive, resistive, or insulating films are deposited or sputtered on a ceramic or other insulating substrate. The films can be deposited either in a required pattern or as a complete film and photoprocessed and etched to form the pattern.

The term thin film is derived from the fact that the deposited films are of the order of a few micrometers in thickness compared with the 10 to 50 μm for thick film. Figure 1.3 illustrates typical thin-film dimensions. Often, thin-film conductors are plated up to improve conductivity. In this case the plating may be as thick as 0.5 to 1 mil, making the thickness of plated thin-film conductors approximately equal to that of thick-film materials.

2
SUBSTRATE MATERIALS AND PROPERTIES

2.1 SUBSTRATE PROPERTIES

The *substrate* is the mechanical base and electrically insulating material used for hybrid fabrication. The material properties of the substrates affect both the processes used and final characteristics of the hybrid circuit.

The property considerations of a potential substrate material should include the following:

Engineering Concerns

1. *Dielectric constant*: Capacitance effects are directly proportional to dielectric constant.
2. *Dielectric strength*: This determines the voltage breakdown properties of the substrate.
3. *Dissipation factor*: This determines the electrical losses in the substrate, and may be particularly important in high-frequency and microwave circuits.
4. *Thermal conductivity*: This determines thermal properties of the circuit, i.e., how much heat film components or added discrete components can dissipate without excessive increases in temperature.
5. *Thermal coefficient of expansion*: This is a major factor in determining the compatibility between the substrate and the materials used in fabrication of the hybrid. It has a significant effect on the film adhesion and the temperature coefficients of resistivity and capacitance.
6. *Volume Resistivity*: Determines the electrical insulation between circuit elements on the substrate.

Manufacturing Concerns

1. *Ability to withstand high temperatures*: Typically 500 to 1000°C for most circuits.
2. *Mechanical strength*: Important for ease of handling.
3. *Surface finish*: Important for good line definition and printed film uniformity. However, in thick film the surface finish should be sufficiently rough to ensure adequate adhesion of the fired films.
4. *Camber*: There should be a minimum of distortion or bowing of the substrate. Too much camber can result in screening problems in thick film and photoprocessing problems in thin film.
5. *Visual defects*: Surface defects such as small pits or burrs can result in circuit defects such as open circuits and pinholes.
6. *Materials compatibility*: The substrate should be chemically and physically compatible with the chemicals and materials used in fabrication.
7. *Low cost in quantity production*.
8. *Tolerances*: Should be possible to achieve substrate tolerances both in overall dimensions and in positioning of holes in the substrate.

2.2 SUBSTRATE MATERIALS

The majority of substrates used in thick-film hybrid fabrication are produced from ceramic materials such as alumina, beryllia, magnesia, thoria, zirconia, or combinations of these materials. The industrial standard is alumina, which offers excellent combinations of the electrical and mechanical properties required.

Thin-film circuitry requires substrates with a smoother surface finish than those used for thick film, resulting in the use of very highly polished ceramic, glass, quartz, and sapphire for these circuits. Quartz and sapphire are often used in microwave circuitry because of their excellent high-frequency electrical properties. Table 2.1 shows the properties of typical thick-film substrates and Table 2.2 the properties of typical thin-film substrates.

Alumina substrates used for thick film are usually in the range 94 to 96%. The grain size is approximately 3 to 5 μm and the surrounding glassy grain boundaries react with the thick-film binder glasses to give significantly higher bond strength compared with other substrate materials.

Steatite and fosterite have sometimes been used for thick-film circuits. They have a lower dielectric constant than alumina (6.1 and 6.4, respectively, compared with 9.0 for 96% alumina), but have lower

TABLE 2.1. Properties of Thick-Film Substrates

Characteristic	Unit	Conditions	96% Alumina	99.5% Alumina	99.5% Beryllia
Dielectric constant		1 MHz	9.3	9.9	6.9
		1 GHz	9.2	9.8	6.8
Dielectric strength	V/mil		210	220	230
Dissipation factor		1 MHz	0.0003	0.0001	0.0002
		1 GHz	0.0009	0.0004	0.0003
Thermal conductivity	W/cm-K	25°C	0.351	0.367	2.5
		300°C	0.171	0.187	1.21
Thermal coefficient of expansion	°C	25–300°C	6.4×10^{-6}	6.6×10^{-6}	7.5×10^{-6}
Bulk resistivity	Ω/cm	25°C	10^{14}	10^{14}	10^{14}
		100°C	2×10^{13}	7.3×10^{13}	10^{14}
Tensile strength	psi		25,000	28,000	23,000
Surface finish	μin.		25–40	10	20
Camber	mil/in.		4	4	3

TABLE 2.2. Properties of Thin-Film Substrates

Characteristic	Unit	Conditions	Alumina	Tape cast beryllia	Corning 7059 glass	Fused silica (quartz)	Sapphire
Dielectric constant		1 MHz	10.1 at 1 GHz	6.9 at 1 MHz	5.84 at 1 GHz	3.826 at 1 MHz	9.39 at 1 GHz
		8.6 GHz	10.7 at 9.9 GHz	6.80 at 1 GHz	5.74 at 8.6 GHz	3.824 at 6 GHz	9.39 at 10 GHz
Dielectric strength	V/mil		770	230		410	190
Dissipation factor		1 MHz	0.0002	0.0002	0.0001	0.000015	0.0001
		8.6 GHz		0.0003	0.0036	0.00012	0.00005
Thermal conductivity	W/cm-K	25°C	0.367	2.5	0.017	0.014	0.417
		300°C	0.187	1.21	0.008	0.008	
Thermal coefficient of expansion		25°C	6.7×10^{-6}	7.5×10^{-6}	4.6×10^{-6}	0.49×10^{-6}	
Bulk resistivity	Ω/cm	25°C	3.16×10^{10}	10^{14}	10^{14}	3.16×10^{10}	10^{14}
Thickness available	mil		10–40	25	32 & 64	Specify	Specify
Tensile strength	psi			23,000		7000	58,000
Surface finish	μin.		1.0	15–20	1	1	1
Camber	mil/in.		3	3	1	1	1

Substrate Materials and Properties

tensile strength and thermal conductivity. However, most thick-film materials and processes are incompatible with these substrates and they have been used only where the lower dielectric constant and usually lower cost have overcome the other disadvantages.

When power dissipation is a problem, beryllia (BeO), with its high thermal conductivity, can be used to advantage. The thermal conductivity of BeO is approximately eight times that of alumina. However, BeO is somewhat weaker than alumina and more expensive to produce in substrate form. Also, beryllia is highly toxic in either powder or vapor form, and requires special handling and safety equipment when machining or firing at high temperatures.

Porcelain steel is becoming popular as a thick-film substrate. The term refers to a low-carbon-steel substrate coated with an electronic-grade glaze about 5 mil thick. The steel is coated with a low-alkali-content glass frit and fired at about 850°C. At this temperature the glass flows to produce a smooth glazed surface called porcelain steel. The substrates can be refired up to about 625°C without damage and require special thick-film pastes that fire below this temperature. Conductor, resistor, and dielectric pastes have been developed which are compatible with porcelain steel. The advantages of this material are its high mechanical strength, easy machinability, excellent thermal properties, and low cost. Capacitance coupling to the steel substrate through the 5-mil porcelain coating is often a major disadvantage at high frequencies.

Higher-purity alumina substrates (99.9% purity) with highly polished or glazed surfaces are the most common thin-film substrates. Glass substrates are also used, particularly in low-power, low-frequency, high-precision circuitry. However, glass is a very poor thermal conductor compared with alumina and care should be taken if the glass substrate is to dissipate any significant power. Tape-cast beryllia substrates are also used in thin-film high-power applications. Quartz and sapphire substrates have extremely low loss compared with other substrates and have found extensive use in microwave thin-film circuits. Sapphire has much higher thermal conductivity than quartz and provides a good thin-film substrate for higher-power microwave circuits. However, both sapphire and quartz are very costly and are not machined as easily as alumina.

2.3 ELECTRICAL SUBSTRATE SPECIFICATIONS

Dielectric Constant

The dielectric constant determines the amount of capacitance associated with film elements deposited or screened on the substrate. For example, we may have a substrate with a ground plane on one side and conduc-

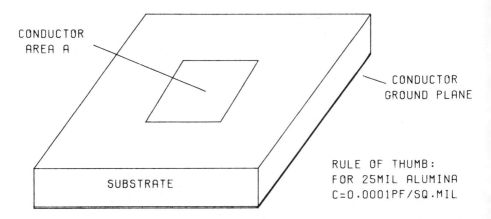

FIGURE 2.1. Capacitance to ground plane.

tors on the other (Figure 2.1). The parallel plate capacitance from the conductor area (A) to the ground plane is given by

$$C = 0.225 \frac{\varepsilon_r A}{t}$$

where

C = capacitance, pF (10^{-12} F)
ε_r = dielectric constant
A = plate area, in.^{2}
t = thickness of the substrate, in.

For a 25-mil-thick alumina substrate with dielectric constant 9.6, this gives a capacitance of approximately 0.0001 pF/mil^2.

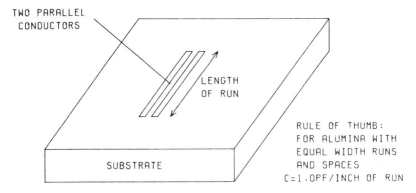

FIGURE 2.2. Capacitance coupling between parallel conductors.

Substrate Materials and Properties

Example A typical thick-film conductor, 10 mil wide × 250 mil long has a 2500-mil^2 area and would have a stray capacitance to ground of 0.25 pF, a fairly high value in most high-frequency circuits.

The substrate dielectric constant will also affect the capacitance coupling between runs on the substrate. This is a more complex problem to solve and is discussed at greater length in Chapter 4. However, a good rule of thumb is that for an alumina substrate with two equal-width runs separated by a gap equal to their width, the capacitance is approximately 1 pF per inch of run (Figure 2.2).

Dielectric Strength

The dielectric strength of an insulator is defined as the voltage gradient (volts per unit distance) at which the material breaks down and starts to conduct. The dielectric strength of insulating ceramic substrates decreases considerably with increasing temperature, frequency, and material thickness. Typical strengths are of the order of 200 rms V/mil for alumina and beryllia substrates. This is usually well above the requirements of normal hybrid circuitry and is not often a major consideration in design.

When using ceramic substrates in high-voltage circuitry, manufacturers should be consulted in detail for information on the dielectric strength properties of the proposed materials.

Dissipation Factor

Dissipation factor is a measure of the electrical loss characteristic of an insulating material and is usually of greatest importance at high frequencies, particularly at microwave frequencies. The dissipation factor is a function of temperature and frequency, increasing with both these variables.

In any solid dielectric there is a power loss due to leakage currents, and also a dielectric heat loss when an alternating voltage is applied across the dielectric. The total loss may be represented as the loss in an equivalent resistor connected across the dielectric. The equivalent circuit of two parallel conductors with the substrate material in between looks like a capacitor and a parallel resistor or a capacitor and series resistor, as shown in Figure 2.3.

For the parallel equivalent circuit the loss current I_r will be very much less than the capacitance current I_c, so that the resultant current I leads the voltage by almost 90° (Figure 2.4).

The difference between 90° and the actual phase angle of the capacitor is the angle δ shown in Figure 2.4. This is termed the *loss angle* of the capacitor or dielectric. For the parallel equivalent circuit,

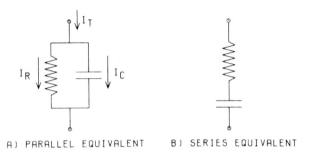

A) PARALLEL EQUIVALENT   B) SERIES EQUIVALENT

FIGURE 2.3. Equivalent circuits of a dielectric.

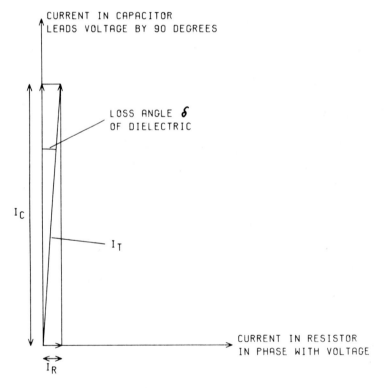

FIGURE 2.4. Loss angle δ of a dielectric.

Substrate Materials and Properties 17

$$\tan \delta = \frac{I_R}{I_C} = \frac{1}{R_p \omega C_p} = \frac{\text{capacitative reactance}}{\text{parallel resistance}}$$

The term "tan δ" is called the *loss tangent or dissipation factor* of the dielectric. For the series equivalent circuit,

$$\tan \delta = R_s \omega C_s = \frac{\text{series resistance}}{\text{capacitative reactance}}$$

The value of tan δ will be the same for both circuits since they are equivalent representations.

The parallel equivalent circuit is usually used at low frequencies and the series equivalent circuit at high frequencies.

Volume Resistivity

The volume resistivity is the resistivity of the bulk ceramic material. Since most substrates used for hybrids are electrical insulators, this is usually a high value in the range 10^{10} to 10^{14} Ω/cm at room temperature.

2.4 MECHANICAL SUBSTRATE SPECIFICATIONS

Dimensional Stability

Alumina ceramics are dimensionally highly stable materials. They have properties of high hardness, rigidity, and low thermal expansion. Within the limits of their tensile and compressive strengths, they are completely rigid and maintain machined tolerances under the high temperatures experienced in hybrid processes.

Alumina ceramics have virtually no plastic deformation and exhibit only small amounts of elastic deformation under high loads. Minute flexing of substrates can occur but the ceramic always returns to its previous shape because it is almost perfectly elastic.

The material also exhibits uniform and predictable thermal expansion characteristics. The thermal expansion coefficient is about 6.4×10^{-6}/°C for alumina, and 7.5×10^{-6}/°C for beryllia.

Substrate Surface Factors

The surface finish of substrates directly affects circuit loss, adhesion of films, and line definition. The substrate surface requirements for thick-film deposition are not as stringent as those for thin-film deposition. Typical thick-film substrates have surface finishes of 20 to 40 µin. [center line average (CLA)]; thin-film substrates, from 1 to 10 µin. In general, substrates should have rough surfaces, in that this gives better adhesion. However, circuit losses increase rapidly with increasing roughness. Improving surface finish from about 8 to 2 μin. can reduce circuit losses by 20 to 30%. This can be significant in high-frequency and microwave circuitry. However, improving surface finish from 8 μin. to 2 μin. may add 25% or so to the cost of the substrate. Surface finish is usually quoted in terms of CLA roughness. Consider a surface finish shown in Figure 2.5.

The center line is drawn over an assigned length of surface so that the areas above the curve equal the areas below the curve.

The definition of roughness (CLA) is then

$$CLA = \frac{A + B + C + D}{L}$$

where

A, B, C, D = areas under or above center line
L = assigned length of substrate

CLA units are usually quoted as μin./in.

It is important to remember that surface flatness and waviness are not included in the surface roughness (CLA) measurement.

Surface Flatness

Surface flatness is often referred to as camber, warpage, or waviness of the substrate. Typically, substrates have a flatness of 1 to 2 mil per inch of substrate length. This can result in major problems in film screening, photoprocessing, and stability of circuits. Nonflat substrates do not press uniformly against screen printers or photomasks, and this usually results in nonuniform ink deposits or etched components. Also, stressing of the substrate when mounting with connectors or in packages results in similar stresses in the film components on the substrate, resistors in particular. Such stresses may change the resistor values in an unpredictable way, and can cause microcracking and even failure in some cases. Thicker substrates reduce the effects of such stresses and should be considered for precision networks.

Substrate Tolerances

Substrate thickness normally ranges from 0.020 to 0.060 in., with 0.025 in. being the industry standard. Thickness tolerances are usually from ±0.001 to 0.004 in., with typical cambers of 0.001 to 0.005 in./in.

Substrate sizes range from about 1/2 × 1/2 in. to about 5 × 5 in. Tolerances on size and hole positioning are typically of the order of 0.002 to 0.004 in. More precise tolerances on length, width, thickness,

Substrate Materials and Properties 19

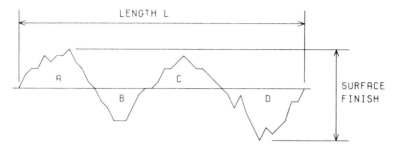

FIGURE 2.5. Surface finish: centerline average.

and camber can be obtained by using a "ground" substrate. To obtain such a substrate, the manufacturer would start with a fired substrate larger than the final dimensions. The surfaces would be ground to within a few mils of the final thickness and camber, and then lapped and polished to the required thickness. The edges are then cut using a diamond dicing saw. Grinding the edges can improve the length and width tolerances even further. Such techniques are obviously expensive but can result in substrates with thickness tolerances of ±0.0005 in. and length and width tolerances of ±0.001 in.

2.5 SUBSTRATE FABRICATION: ALUMINA

Alumina powders with particle sizes of 3 to 50 µin. are blended with various additives such as clay (silica), talc (magnesia), and calcium carbonate (calcia). The silica acts as a glass-forming agent and the magnesia and calcia act as catalyst fluxing agents to reduce the glass-forming temperature. The powders may be ball or roll milled for up to 10 hr with plasticizers, organic binders, lubricants, and solvents, to ensure complete mixing. For dry press fabrication the slurry is spray dried to obtain a uniform particle size.

Dry or Powder Pressing

The dry or powder pressing process is usually used for substrates greater than 0.040 in. thick. The powder is packed into an abrasion-resistant die (carbide) under high pressure (8000 to 20,000 psi) and pressed or compacted. Shrinkage factor in the subsequent firing is about 16% and dies are designed oversize to compensate for this. Holes and various shapes are possible with this process, although holes should not be located too close to an outside edge. The process allows automatic production of up to 3500 pieces per hour with reasonably controlled tolerances.

Sheet Casting: Green Tape

Sheet casting is a low-pressure process and is generally used for parts less than 0.040 in. thick. The oxides are prepared as a slurry by adding organic binders, plasticizers, and solvents. This slurry is allowed to flow out on a long belt of very smooth material, usually Mylar. The film and slurry move under a metal doctor blade positioned above the film. This blade is used to control the resultant thickness. The material is air dried to remove the 25 to 50% solvent content. The resultant sheet is usually referred to as being in the "green" state because of its color. During subsequent firing this material has a high shrinkage factor, usually 18 to 22%. Individual parts can be cut out of this green tape prior to firing. Holes, slots, and other shapes can also be easily cut or punched out. The punches used are generally made of carbide and must compensate for the material shrinkage factor.

Scribed lines approximately 0.005 to 0.010 in. thick can also be made in the green sheet. This allows for multiple processing in the thick-film screening process. The scribe lines set up a fracture plane which facilitates breaking up the main substrate into individual parts after screening. These scribe lines can be added as part of the cutout die or by using ganged slitting blades.

Firing

The individual substrates formed by the processes described above need to be fired to form usable parts. The substrates are stacked on flat kiln plates and fired through two temperature profiles:

1. Prefire at 300 to 600°C to burn off organic binders, lubricants, and plasticizers
2. Sintering at 1500 to 1700°C, which consolidates the fine particles into the final substrate form

A typical firing cycle is 18 to 24 hr with 4 hr at the peak temperatures. After firing the substrates are said to be *as-fired*. They are generally tumbled to remove small burrs and nonadherent contaminants.

Glazing

As-fired substrates are usually suitable for thick-film processing. However, these surfaces are usually too rough for thin-film processing. The as-fired surfaces can be improved by coating one or both surfaces with thin glassy layers called glazes. Low-melting silicate glasses are applied to the as-fired substrate, which is then refired at lower temperatures. The glaze melts and fuses to the substrate surface, giving an improved surface finish.

Laser Scribing

High-powered CO_2 laser systems are finding significant usage for scribing and machining holes in alumina and beryllia substrates. Due to shrinkage associated with firing the substrates, it is difficult to maintain accurate tolerances and hole coordinates. Laser scribing/ machining takes place after firing and leads to more accurate substrate dimensions and hole locations.

Holes are usually laser drilled prior to thick- or thin-film processing, whereas scribe lines are usually added after film processing.

3
THICK-FILM MATERIALS

3.1 INTRODUCTION

There are three main categories of thick-film pastes or inks: conductors, resistors, and dielectrics. Each paste has three main constituents: the binder—glass frit; the vehicle—organic solvents and plasticizers; and the functional elements—metals, alloys, oxides, or other ceramic compounds. The relative proportions of binder and functional elements determine if the end result is a conductor, dielectric, or resistor.

3.2 THICK-FILM CONDUCTORS

Conductor materials are usually used in the greatest quantity in any thick-film operation and usually contribute the highest material cost.

Functions

The principal functions of conductors are as follows:

1. To interconnect components
2. To terminate resistors
3. To serve as attachment pads for discrete components
4. To serve as thick-film capacitor electrodes

Material Properties

The following material properties are required:

1. Good conductivity, 0.002 to 0.15 Ω per square
2. Good adhesion, 2000 psi for 100-mil^2 pad
3. Provide good wire bondability
4. Provide good eutectic die attachment
5. Solderable with high leaching resistance

6. Stable during processing
7. Good line definition and resolution
8. Good screening properties with no ink runoff, smearing, or scalloping
9. Long shelf life

Conductor Formulations

Binder

The binder consists of low-melting-point glasses which hold the metal particles in contact and bond the film to the substrate. Properties of interest when manufacturers formulate the materials are glass temperature/viscosity relationships, surface tension properties, chemical reactivity, and coefficient of thermal expansion.

The viscosity of glass decreases steadily with increasing temperature, and the temperature flow characteristics must be properly controlled for the glass to wet or coat the metal particles sufficiently to hold them and to wet the substrate. However, the metal particles should not become isolated by this wetting process, as this would result in poor conductivity. Also, the glass should not become too fluid and flow out over the substrate, leaving a porous, poorly defined pattern. The chemical reactivity of the glass determines the quality of the bond to the substrate. The thermal coefficient of expansion of the binder glass should match that of the substrate reasonably well. Typically, the binder is about 10 to 20% by weight of the conductor formulation.

Vehicle

The vehicle helps define the printing characteristics of the material. The vehicle usually contributes about 12 to 25% by weight of the conductor paste. Typical solvents used include alcohols, pine oil, and α-terpineol.

Conductive Material

The conductive materials are small metallic particles less than 5 μm in size. Typically, the conductive metal is 50 to 70% by weight of the conductor paste. The particle size, size distribution, and particle shape have a significant influence on the electrical and physical properties of the final film conductor. In general, precious metals such as gold, silver, palladium, and platinum are used as the conductive materials since other metals usually oxidize in high-temperature firing processes.

Conductor Types

Silver (Ag)

Silver provides the lowest-cost material for conductors. It also has the properties of good substrate adhesion, good solderability, and ease of processing.

However, silver pastes find very little use due to the well known tendency for silver to migrate under the influence of dc fields. This results in the silver reacting with most resistor films. Silver also has poor solder leach resistance, and is also susceptible to tarnishing.

Gold (Au)

Gold finds wide usage as conductor material for thick film. It is almost always used when semiconductor dice are to be eutectically bonded to the hybrid. It also provides excellent wire bonding characteristics and low resistance (0.003 to 0.01 Ω per square). However, gold pastes are very costly and may have a large price fluctuation as the price of gold fluctuates. Gold also suffers from the property of poor solderability.

Palladium Silver (PdAg)

This material has the highest usage of any material for thick-film conductors, meeting approximately 80 to 90% of the demand for thick-film conductors.

Palladium silver has most of the required properties, such as good substrate adhesion, good solderability, and good solder leach resistance. The material is also only a fraction of the cost of gold-based pastes. However, because of the silver content, the material has a slight silver migration problem. The sheet resistivity of the material is higher than gold, being approximately 0.03 Ω per square. A major disadvantage of the material is that is is difficult to wire-bond to, giving mediocre results at best.

Palladium Gold (PdAu) and Platinum Gold (PtAu)

The addition of palladium or platinum in addition to gold has resulted in these fairly popular thick-film pastes. Their major advantage over gold is that they have good solderability and retain reasonable wire bonding characteristics. The material cost, however, is approximately the same as that of the gold pastes and much higher than the palladium silver. Sheet resistivity is also higher than the gold or palladium silver pastes, being approximately 0.04 to 0.08 Ω per square.

Copper and Nickel

Significant research has been done on replacing the noble metal conductor pastes with low-cost copper- or nickel-based pastes. Such pastes often suffer from oxidation during firing and to prevent this are normally fired in a nitrogen atmosphere. Special furnaces purged with nitrogen gas have been developed for this purpose. Advances are constantly being made with these compositions, and some copper/nickel pastes recently developed can be air fired. Because of the recent large fluctuations in noble metal prices there is a definite trend toward these copper- and nickel-based pastes. In the future these pastes could account for a major share of the thick-film conductor market.

Table 3.1 compares the properties of the most commonly used conductor pastes, and Table 3.2 gives a summary of areas of application.

General Comments on Conductor Pastes

A proliferation of conductor materials are available on the market, generally formulated to achieve a particular property. For example, variations in the amount of palladium in a palladium silver paste can increase leach resistance, as the palladium increases, but solderability and cost will suffer. It is important to remember that there is no such thing as a perfect thick-film paste—the selection process is always a compromise.

The price of the material used follows the metal markets fairly closely and is usually adjusted to the prevailing metal market at the time of manufacture.

TABLE 3.1. Comparison of Conductor Characteristics

	Gold	Palladium silver	Palladium gold	Platinum gold
Typical sheet resistivity (Ω/square)	0.005	0.03	0.05	0.05
Solderability	No	Excellent	Excellent	Excellent
Wire bondability	Excellent	Poor	Good	Good
Die bondability	Excellent	No	Good	Good
Cost	High	Low	High	High

TABLE 3.2. Most Common Application of Thick-Film Pastes[a]

Material	Most common areas of application
Gold	All except 9, 13, 14
Palladium silver	All except 9, 10, 11
Palladium gold	All except 1, 10, 11, 14
Platinum gold	All except 1, 10, 11, 14

[a]Application area codes:

1. Microwave stripline
2. Resistor termination
3. Bottom conductor layer
4. Any conductor layer
5. Multilayer conductor
6. Crossover conductor
7. Bottom capacitor plate
8. Top capacitor plate
9. Switch contact
10. Wire bond
11. Die bond
12. Ground plane
13. Solderable
14. Very low cost

It is important to consider coverage per unit weight when determining the relative economics of different materials. Various conductor materials screened with the same mesh may cover areas varying from 10 to 20 in.^{2}/g. This obviously has a significant cost impact.

Newer compositions gaining popularity are copper-based pastes. These pastes offer very high conductivity and good solderability at a fraction of the price of the lowest-cost palladium silver pastes. Improvements are continually being made in the properties of these pastes.

Multilayer Conductors

Over the years the complexity of hybrids has been increased to the point where several conductor levels are required to interconnect the circuit. This has been true particularly of multichip digital circuits. Simple dielectric crossovers no longer suffice for such complex circuits, and sophisticated multilayer structures containing several conductor and dielectric layers are required.

The conductors used in multilayer hybrids require good adhesion to both substrate and fired dielectrics, high electrical conductivity, and dense, smooth surfaces for reliable wire bonding. The dielectrics need thermal expansion coefficients matched to the substrate material, low dielectric constant, and high insulation resistance.

The usual process sequence for fabricating a multilayer structure is as follows:

Print and fire first conductor.
Print and fire first dielectric layer with vias (holes in dielectric) for connections from second conductor level to first conductor level.
Print conductor to fill up vias in the dielectric.
Print second conductor and fire.

This sequence is repeated for the number of conductor layers required. Hybrids with up to seven levels of conductors are becoming commercially available.

Because of the need for increasingly complex, highly dense digital networks, multilayer hybrids may become the packaging technology of the future.

3.3 THICK-FILM DIELECTRICS

Functions

Dielectrics are used as crossover insulators for multilayer conductors, capacitor dielectrics for thick-film capacitors, and as an overall encapsulation glaze.

The capacitance associated with crossovers is usually an undesired characteristic. Thus crossover dielectrics are designed with low-dielectric-constant materials. Capacitor dielectrics, however, are doped with various amounts of barium titanate to increase the dielectric constant.

Material Properties

The following material properties are required:

1. Breakdown strength of 250 V/mil
2. Insulation resistance greater than 10^{10} Ω
3. Dissipation factor less than 0.1% for crossovers
4. Dielectric constant 6 to 8 for crossovers, 10 to 2000 for capacitors
5. Minimum tendency to form pinholes
6. High resistance to thermal shock crazing
7. Good screening properties
8. Long shelf life

Dielectric Formulations

Binder

The requirements for the glass system in dielectric pastes are fairly severe since the dielectric has to provide some important properties. The film must be compatible with substrate and conductor materials.

To avoid short-circuit problems, the film must also be pinhole free. The dielectric film should also have minimum flow and softening when used as a crossover.

To achieve a dense pinhole-free film, the material needs to flow reasonably well. To achieve this, a low-viscosity glass binder is required. However, in contradiction to this, a low-viscosity material could allow a top conductor to sink or swim in the material during firing. This could result in possible shorts in a crossover application or, at best, registration errors between top and bottom conductors. To avoid this phenomenon a high-viscosity glass binder is required.

Manufacturers have resolved this problem by carefully balancing the viscosity or utilizing crystalline glass formulations which result in a higher-viscosity material after initial firing operation. A 10:1 change in viscosity is possible on subsequent refiring.

Vehicle

The vehicle used in dielectric pastes is generally similar to that in the conductor pastes.

Dielectric Materials

The characteristics of the fired capacitor depend on the nature of the dielectric material incorporated in the paste. There are two classes of materials, those with low permittivity and those with high permittivity.

Dielectric Types

Materials with high dielectric constants, called high-K types, are based largely on the ferroelectric ceramic, barium titanate. This material has a dielectric constant of about 1200 at room temperature. Addition of strontium, calcium, tin, and/or zirconium oxides to the barium titanate reduces the temperature coefficient of capacitance to reasonable values. Compositions with dielectric constants in the range 1000 to 3000 with temperature coefficients of ±5000 ppm/°C have been formulated.

Several low-permittivity, nonferroelectric dielectric materials based on magnesium titanate, zinc titanate, titanium oxide, and calcium titanate are available. Dielectric constants are in the range 12 to 160, with temperature coefficients of ±200 ppm/°C. By varying the proportions of magnesium and/or zinc oxide to titanium oxide, compositions are possible having a NPO characteristic. [Negative positive zero (NPO) means that the temperature coefficient goes from a negative value to a positive value over the temperature range of interest.] High-K dielectrics are not as stable as low-K dielectrics. High-K dielectrics exhibit a slow reduction in capacitance with time. The change tends to be most rapid in the period immediately after firing, with relatively slow aging thereafter. The changes may be of the order of several percent. Low-K dielectrics are significantly more stable than higher-K types and should be used in cases where capacitance drift is a critical parameter.

For crossovers and multilayer circuits the need is for a dielectric with a low dielectric constant (low-capacitive coupling between conductors). Glass/ceramic mixtures have been developed which are well suited to these applications. These materials have low dielectric constants (10 to 40) and exhibit the required increase in viscosity after the initial firing operation. After first firing at temperatures generally close to 850°C, these compositions can be refired to approximately 1000°C without appreciable loss of dimensional stability.

The final dielectric material used is for the purpose of glass encapsulation of the final circuit. These encapsulant glaze compositions provide environmental protection for printed resistors and capacitors. These compositions usually comprise vitreous materials whose melting points are much lower than any of the other thick-film pastes used. Firing temperatures are normally in the range 450 to 500°C. Metal oxide pigment is often added to the paste to give it certain colors.

3.4 THICK-FILM RESISTORS

Resistor pastes are second to conductors as far as usage is concerned. The resistor pastes can be obtained in various sheet resistivities from approximately 1 Ω per square to 10 MΩ per square. This wide variance in available resistor values enables almost any analog or digital circuit to be "hybridized" using thick film.

Material Properties

Material properties required include the following:
1. Wide range of resistor values
2. Compatible with substrate and termination conductor materials
3. Low-temperature coefficient of resistance (TCR)
4. Good TCR tracking between resistors
5. Low-voltage coefficient of resistance
6. Material should exhibit little drift at elevated temperature, during thermal cycling, under humid conditions, or over its normal load life. The overall stability should be in the change range 0.1 to 0.5%
7. Low current noise due to imperfections, hot spots, and general nonhomogeneity of the thick-film material
8. Capable of being blended to form intermediate sheet resistivities.

Resistor Formulations

Binder

For resistor formulations the glass selection is more critical than in conductors, since the glass systems have a significant effect on the resistor properties. Factors such as particle size and firing conditions can affect the resistivity and TCR of the final film. Composition of the glasses used varies from manufacturer to manufacturer, with lead bismuth borosilicate ($PbBiBSiO_2$), lead bismuth, and lead zirconate sometimes being used in conjunction with various oxides. The glass frit material occupies about 40% of the composition by weight.

Vehicle

The resistor vehicle constitutes about one-third of the composition by weight and defines the printing characteristics of the materials. Typical resin materials used are ethyl cellulose together with solvents such as terpineol.

Conductive Material

The conductive materials used to form the resistor paste are usually metals, metal oxides, or a combination of these, and occupy about 25% of the composition by weight. Generally, several different metals are used to achieve the particular properties desired. The theory of what actually creates the conduction is not fully understood, although it is recognized that it is related to semiconductor conduction mechanisms.

Resistor Materials

The first commercially available thick-film resistor pastes were based on compositions of palladium, palladium oxide, and silver. Pastes formulated from these materials were found to be very sensitive to the firing profile used and have been superseded by more advanced systems based on ruthenium, iridium, and rhenium. These systems offer significant improvements, particularly as far as TCR is concerned, and are much less process sensitive. Typical firing temperatures are in the range 700 to 900°C, depending on the composition.

Ruthenium oxide is now probably the most common material used in thick-film resistor pastes. The oxide is very stable, both toward oxidation and reduction at elevated temperatures and may be heated to about 1000°C in air without undergoing chemical change. Formulations using ruthenium oxide and lead glass give pastes with a wide range of sheet resistivity, good stability, relatively low values of TCR, and low electrical noise. Since the resistive component is stable at elevated temperatures, no chemical reaction takes place during firing,

Thick-Film Materials

and this gives the material a good tolerance to variations in the firing profile. The fired films are also relatively insensitive to subsequent high-temperature processing steps. The ruthenium oxide films also exhibit long-term thermal and load stability characteristics.

The costs of the different types of resistor composition vary widely. The ruthenium oxide pastes are more expensive than those based on palladium-palladium oxide-silver types. However, it must be remembered that 1 troy ounce of paste may cover an area of 400 to 500 in.^{2} when screened. This results in a cost of maybe 5 cents/in.^{2} of paste, and for a normal thick-film circuit the resistor material cost is usually less than a few cents. There is therefore very little saving in choosing a low-cost paste over an improved expensive paste.

3.5 THICK-FILM PROCESSING

Screening

Selective deposition of the thick-film materials is achieved by the use of stencil screens or masks. The process depends essentially on the forcing of ink through a stencil pattern comprising a fine mesh, usually of stainless steel. The process is almost identical to the silk screening process used widely in the production of decorative finishes, labeling of bottles and containers, and in the production of etched or printed circuit boards. However, for the tolerances required in thick-film circuitry the screening equipment has been improved considerably. The key to effective production of thick-film circuits is in the understanding and control of the screening operation.

Screens

Almost all hybrid circuits are produced using stainless steel mesh screens. For very high precision circuits with line widths of 0.002 to 0.004 in., etched metal masks have been used. Standard mesh emulsion screens allow printing of line widths of 0.008 in. with 0.008-in. spacing. The lower limit with the best screens, processes, and materials is about 0.005 in. wide with 0.005-in. spacing. The common mesh types utilized for thick films are shown in Table 3.3.

Standard screens are usually 5 × 5 in. and 8 × 10 in. aluminum screen frames with mesh mounted mechanically or with adhesion after stretching. Typical cost of the mesh is about $2 to $6 per square foot and final screen cost will be from $6 to $30, depending on the material used and the size of the final screen.

Tension control is fairly important in fabrication and can be monitored by applying a given load to the center of the screen and measuring the deflection. The deflection is a function of the mesh type and screen size. Screens can be purchased from specialist manufacturers, often

TABLE 3.3. Common Mesh Types

Mesh count	Wire diameter (in.)	Open area (%)
105	0.003	46.6
150	0.0026	37.8
165	0.0020	45.8
180	0.0020	42.0
200	0.0020	36.0
325	0.0012	38.1

with the pattern already defined. Although initial screen tension is then the responsibility of the manufacturer, it is important to monitor the tension during the life of the screen to prevent yield problems during processing.

Emulsion Screens

The screens are evenly coated with a photographic emulsion and dried. The pattern is then defined by exposing the emulsion to an ultraviolet-light source through a photographic mask to harden areas to be retained on the screen. After exposure, the nonhardened areas, corresponding to the pattern to be printed, are washed away with a developer solution.

Often, the pattern will be applied to the screen at a 22 or 45° angle with respect to the mesh. This is done due to the difficulty encountered in processing of aligning the pattern perfectly with respect to the mesh orientation. This technique is particularly useful in the case of fine-line printing.

Screen Printing

The basic functions of the printer are to provide a mechanism for mounting the screen, a method of holding the substrate, some method of aligning the pattern with respect to the substrate, and a mechanism to adjust the vertical height between the screen and the substrate. They also provide a mount for a squeegee and some method of traversing the squeegee across the pattern at some controllable rate. The screening process provides pressure on the material via the squeegee and forces it through the open pattern areas. The amount of material depends on the rheology properties of the material, the force exerted by the squeegee, and the thickness of the mesh.

The squeegee is generally polyurethane, with a sharp edge usually 45° to the vertical. The squeegee forms a seal with the screen to transfer the material across the screen pattern. The deformation of the squeegee should be constant so that the force or pressure on the material is constant across the printing stroke.

The substrate is positioned a short distance below the screen and a small quantity of thick-film paste is dispensed onto the upper surface of the screen. The squeegee then travels along the screen surface, deflecting it into contact with the substrate and forcing the paste through the open mesh areas of the screen. The screen regains its natural position after passage of the squeegee, leaving the printed paste on the substrate. The substrate is then removed and the process repeated. Substrate registration is usually by three-pin location, and a screening alignment tolerance of about 5 mil between successive screen operations is usually possible.

The screen printing process is probably the most difficult operation for the successful production of thick-film circuits. As many as 20 to 30 variables have been identified, many of which are difficult to control. The following major variables must be understood and controlled to achieve a high-yielding printing process.

1. Rheology properties of the materials
2. Mesh type and tension
3. Mesh thickness
4. Squeegee materials
5. Squeegee speed
6. Vertical distance between screen and substrate (snap-off distance)
7. Pattern being printed
8. Cleanliness of process

Screen printers run from $1000 for simple manual models to $20,000 for automatic substrate loading systems. Typical medium-volume production machines are in the range $5000 to $7000.

Firing

Immediately after screening the solvents are removed by air drying or low-temperature heat-assisted drying. The temperature should not exceed about 100°C during this drying operation. After air drying the substrates can be fairly easily handled, as the dry paste is reasonably resistant to scratches.

After drying, the substrates are fired at temperatures from 500 to 1000°C, depending on the materials being fired. The substrates are normally fired in conveyor belt kilns. These kilns have several zones of controllable heat, which permit a temperature versus time profile to be set up for a particular material. Such a profile is shown in Figure 3.1 and is usually set up to an accuracy of about ±5°C.

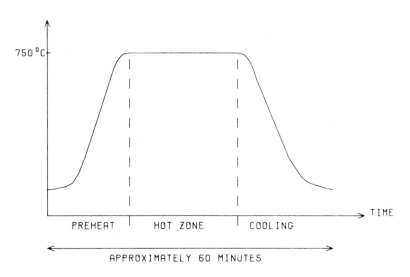

FIGURE 3.1. Firing profile for typical thick-film paste.

Commercial kilns are usually constructed with four or more zones of controllable temperature, exhaust, and atmospheric control. The conveyor belt varies from about 4 in. to 12 in., depending on the production volumes required, and the belt is usually made of Nichrome or equivalent wire mesh. Commercial kilns may vary from $6000 for 4-in.-belt low-volume models to over $50,000 for 12-in.-belt high-volume models.

The parts are loaded on the wire belt and passed through the kiln at a controlled rate. The preheat section burns off any remaining organic constituents and is usually at 250 to 350°C. Appropriate exhaust control should prevent any of the burned-off contaminants from entering the hot zone—where the materials are sintered at temperatures from 500 to 1000°C, depending on the materials being fired. During this sintering process the glass frit melts and wets the substrate, and the glass/metal matrix forms the functional properties of the final film. The cooling zone allows the substrates to return to room temperature with minimal thermal stressing of the films and substrate. The typical profile is usually about 60 min overall with 10 to 12 min at peak temperature in the hot zone.

The kiln temperature profile is very critical in achieving a high-yielding process. This is particularly important with resistors which need to be fired close to their specified sheet resistivity. Production kilns are generally kept loaded with dummy substrates during down-times in order to maintain a constant thermal loading. Temperature variation of the hot zone may affect some resistor films as much as 3%/°C.

Due to sensitivity of the resistor materials to the temperature profile and the difficulty of maintaining a stable profile, it is preferable not to change a single kiln to the various profiles required. Usually, metalization and dielectric films, being less sensitive to temperature variations, can share a single kiln, provided that production loading is such that adequate profile transition time is available. Resistor pastes with different sheet resistivities may have different profiles and may require separate kilns. However, most film manufacturers supply a particular series of pastes with a wide range of sheet resistivity and requiring only one temperature profile. This reduces the requirement to one kiln for the resistors and one for the metals and dielectrics.

Typical firing temperatures for the various thick film materials are as follows:

Material	Firing temperature (°C)
Metallization	900-1000
Dielectric crossover/capacitors	850-1000
Resistors	750-950
Dielectric encapsulation	500-750

The material sequence is selected so that the highest-firing-temperature material is applied and fired first, with each succeeding screening and firing at a lower temperature. This is done to minimize interaction between the materials, and to minimize variations in the properties of the earliest fired materials.

Process Flow

A typical process for a thick-film hybrid comprising metal conductors on the front side, electrical ground plane metal on the back side, a crossover dielectric on the front side, crossover metal on the front side, three different sheet resistivities, and a final encapsulation dielectric would be:

    Screen metal front side
    Air dry
    Screen ground plane reverse side
    Fire at 950°C
    Screen crossover dielectric
    Air dry
    Screen crossover metal
    Air dry

Fire at 900°C
Screen resistor 1
Air dry
Screen resistor 2
Air dry
Screen resistor 3
Air dry
Fire at 850°C
Screen encapsulation
Air dry
Fire at 550°C

Although such a sequence may seem complex to non-process-oriented people, it should be pointed out that many parts can be processed at each step, resulting in a low-cost process.

4
THICK-FILM DESIGN GUIDELINES

4.1 GENERAL DESIGN PROCEDURES

Various functional and environmental requirements will dictate the approach to designing a particular hybrid thick-film circuit. The transition from an electrical schematic to a thick-film circuit can be a complex process. Apart from the "black box" specifications of the circuit, the designer must keep in mind constraints such as size, weight, thermal requirements, cost, and reliability. Also, the design must be compatible with the design rules of available thick-film elements. There needs to be close proximity, coordination, and communication between the design engineer and the process and material engineers. The designer needs knowledge of both the thick-film material properties and limitations.

4.2 SCREENING REFERENCE CORNER AND SCREENING PROCEDURE

Screening Alignment

When screening a thick-film part, a printing fixture with three adjustable pins is usually used to align the part, as shown in Figure 4.1.

The corner lying adjacent to the pins, which is usually the lower right-hand corner of a front pattern, is the reference corner. When a back pattern is also used, the opposite corner on the width becomes the reference corner as the part is flipped over for the screening of the back.

Three tick marks placed on centerlines, oriented as illustrated in Figure 4.2, should be indicated on the process drawing for both front and back patterns. The tick marks have several functions, one being to orient the back-to-front patterns. They are also placed on final artwork and used when making screens to align the pattern square

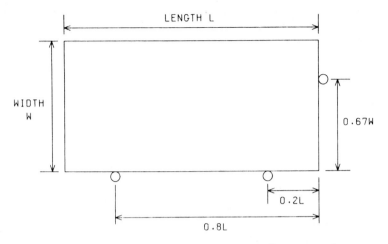

FIGURE 4.1. Alignment fixture for thick-film screening.

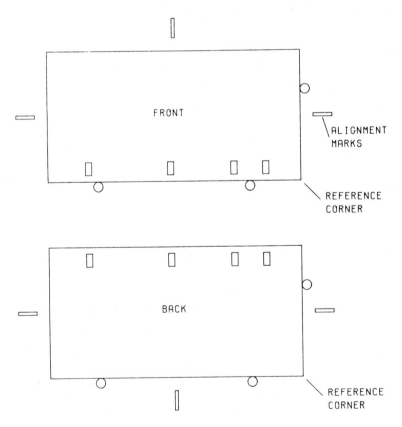

FIGURE 4.2. Alignment marks for screening.

Thick-Film Design Guidelines

with the screen mesh. People generating the final artwork also use the tick marks to place title blocks in the proper position. Consistently placing the title blocks in the same relationship to the reference corner serves as a ready indicator to production of which corner should be used for reference.

The screen is aligned to the reference corner as accurately as possible for printing. Successive printings can be aligned to ±0.005 in. relative to each other in production. With some difficulty, successive printings can be aligned to ±0.002 in., although this will certainly increase part cost.

Margin Consideration

In the printing operation there will be a part-to-part alignment error of about ±5 mil caused by inability to position a part exactly the same in the printer each time. Closer alignment is possible but will add to the cost of the part and is not recommended. To compensate for this variation, a margin or border must be left around the outside of the conductor pattern. The amount of margin necessary is a function of the part size, and typical guidelines are shown in Figure 4.3.

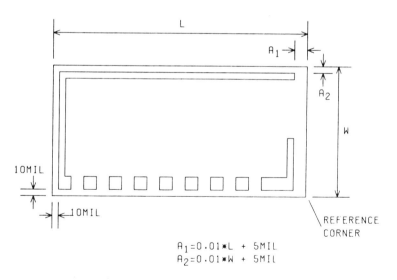

$A_1 = 0.01*L + 5 MIL$
$A_2 = 0.01*W + 5 MIL$

FIGURE 4.3. Layout margin requirements.

4.3 BASIC DESIGN RULES FOR THICK-FILM RESISTORS

Description of Thick-Film Resistors

As described in Chapter 3, the resistors are obtained by screening and firing resistive pastes on to a ceramic substrate. The resistor shown in Figure 4.4 can be considered as a conductive block of uniform thickness T, width W, and length L.

If the bulk resistivity (resistance per unit volume) of the material is ρ (Ω/cm), then the resistance R measured across the block in direction of the length L is given by

$$R = \frac{\text{bulk resistivity} \times \text{length}}{\text{cross-sectional area}}$$

$$= \frac{\rho L}{TW}$$

If ρ/T in this expression is now defined as the sheet resistivity, ρ_s, then

$$R = \rho_s \frac{L}{W}$$

or

$$R = \rho_s \times \text{number of squares}$$

where L/W is the number of squares in the direction in which the resistance is being measured.

The sheet resistivity is defined in ohms per square, and is the basic design parameter specified for film resistors, both thick and thin film.

Example If the resistor is screened with paste of sheet resistivity 1000 Ω per square, and the resistor is defined as 0.200 in. long and 0.040 in. wide, then the resistance R is given by

$$R = 1000 \frac{\Omega}{\text{square}} \times \frac{0.200 \text{ in.}}{0.040 \text{ in.}} \text{ squares} = 5000 \text{ }\Omega$$

It is important to note that for any given film of constant thickness, increasing the length and width by equal ratios will not alter the resistance value. For example, a 40 × 40 mil resistor has the same value as a 100 × 100 mil resistor. They both have the dimensions of 1 square. The sheet resistivity of a thick-film paste is determined by screening and measuring a resistor of unit length and width. Resistor paste manufacturers specify sheet resistivity for a specified thickness T, and obtain different sheet resistivities by altering the resistive paste material composition. Most resistive pastes are specified at about 0.8 to 1.0 mil thickness.

Thick-Film Design Guidelines

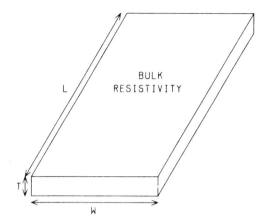

FIGURE 4.4. Sheet resistivity of a thick-film resistor.

Resistor Design Guidelines

The *aspect ratio* of a thick-film resistor is the ratio of the resistor's length to width. This aspect ratio should not be greater than about 20:1 or less than about 1:3. Rather than design resistors with long thin aspects or short wide aspects it is usually more convenient and area saving to change to a different sheet resistivity. For example, rather than designing a 1-MΩ resistor with 100 squares of a 10-kΩ per square paste, it is easier to design it with 1 square of 1-MΩ per square paste.

However, since a separate piece of artwork, a separate screen, and a separate screening operation are required for each resistor paste, it is important to select the inks to minimize their number and optimize the aspect ratios of the resistor. Wherever possible, the number of resistor pastes should be kept to two or three. If resistors are required on both sides of the substrate, care should be taken to partition inks of the same resistivity to the one side. This will again minimize the number of screens and screenings. Other requirements, such as power dissipation, tracking requirements, and voltage stress, may well determine the resistor geometry. Resistor pastes are available with sheet resistivities from 1 Ω per square to 10-MΩ per square in decade steps. Pastes can be blended to give intermediate sheet resistivities if required, although for reasons of economy and inventory control the number of blends should be minimized. The minimum dimensions of resistors are 20 mils wide and 30 mils long, although twice these dimensions is desirable if stability is required. Figure 4.5 shows typical resistor geometries and dimensions for a subsquare and a multisquare resistor. As a rule of thumb the power density should be kept

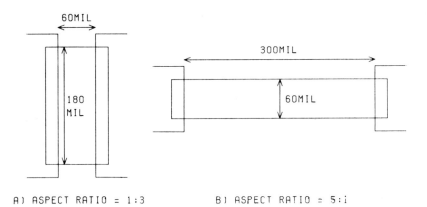

A) ASPECT RATIO = 1:3   B) ASPECT RATIO = 5:1

FIGURE 4.5. Thick-film resistor geometries.

below 50 W/in.^{2} and the voltage stress below 1 V/mil. The total dissipation of resistors plus active devices should not exceed about 1 to 2 W/in.^{2} of substrate area.

If the number of squares required is very large, and for some reason a higher sheet resistivity is not available to reduce the square count, then a zigzag (meander) resistor configuration can be used as shown in Figure 4.6.

When calculating the square count of a zigzag resistor, each corner should be counted as approximately 1/2 square. For example, the zigzag resistor shown in Figure 4.6 would be of 30 squares.

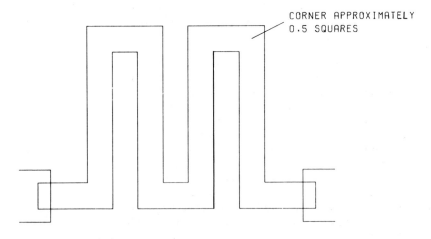

FIGURE 4.6. High-square count resistor geometry (30 squares).

Resistor Tolerances

Due to variances in materials, screened thickness, and resistor definition, the typical as-screened value of a thick-film resistor will be approximately ±20%. Resistors requiring only ±20% tolerance or greater should be designed in the layout stage to nominal value. If resistors are required with less than 20% tolerance, they should be designed in the layout stage to 80% of nominal value or less. The resistors will then need to be sand or laser trimmed to value after firing. The trimming process and procedures are described in detail in Chapter 7.

Resistors will increase in value during trimming and sufficient trim ratio needs to be available in the designed resistor. A trim ratio can be defined as

$$\text{trim ratio} = \frac{\text{number of layout squares after maximum trim}}{\text{number of layout squares before trim}}$$

Example Consider a resistor with required final value of 10 squares designed with a ±20% process. The initial layout would be for 20% less squares, or 8 squares. The process could now come in at minus 20% in the worst case, giving an as-figured value of 6.4 squares. To adjust this resistor back to 10 squares, we would need to add 3.6 squares. But each resistor square added is 20% low. Hence we need a trim capability of 3.6/0.8 squares ≡ 4.5 squares. Thus the

$$\text{trim ratio (20\% process)} = \frac{8 + 4.5}{8} = \frac{12.5}{8}$$

$$= 1.56$$

The resistor would therefore be laid out as 8 squares with 56% × 8 squares = 4.48 squares possible trim variation. Because of the loose tolerances involved, the possible trim variation would be rounded up to 4.5 or even 5 squares.

Resistor Trim Geometries

To achieve the trim ratios required, various resistor trim geometries and trim cuts are available. Some of the popular trim configurations are shown in Figure 4.7. Care must be taken in the layout stage such that after trimming the resistor does not violate any of the basic design guidelines, such as 20-mil minimum material width. Violation of these guidelines may lead to hot spots due to current crowding in the resistor, with loss of stability and possible resistor failure. With this in mind it should be obvious that a top hat, for example, should be at least 40 mil wide. For ease of trimming resistors should, where possible, be oriented in the X-Y direction.

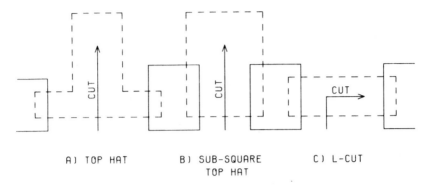

A) TOP HAT   B) SUB-SQUARE TOP HAT   C) L-CUT

FIGURE 4.7. Popular resistor trim geometries.

The laser and sand trimming processes remove the resistor down to the ceramic substrate, forming a clean cut or kerf. Using these methods, stable resistors with tolerances better than ±0.25% are possible.

Resistor Terminations

The resistors are terminated with conductor pads and the resistor should overlap the conductor pad by at least 20 mil. The metallization should extend a minimum of 5 mil beyond the resistor termination perpendicular to the current flow. Figure 4.8 shows a well-designed resistor termination.

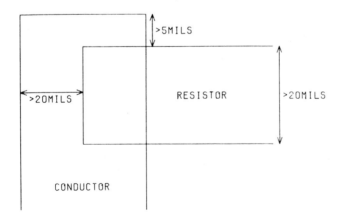

FIGURE 4.8. Thick-film resistor terminations.

Thick-Film Design Guidelines 45

Other Resistor Layout Guidelines

1. The side of the resistor that will not be trimmed should be located at least 20 mil from any other screened area.
2. The trimmed side of the resistor should be located at least 40 mil from any other screened area.
3. If resistor pastes with different sheet resistivities terminate on a common pad, they should be separated by at least 10 mil, as shown in Figure 4.9.
4. Resistors of the same resistivity terminating on a common pad may form a continuous film as shown in Figure 4.10.
5. When a resistor termination pad also serves as a contact pad for a terminal or discrete component, the termination must be designed such that bleed-out of the resistor during firing does not affect the component add-on area of the termination. Figure 4.11 shows typical configurations often used.
6. To maintain optimum as-fired tolerances, and for ease of trimming all resistors should be oriented in the X-Y direction.
7. All resistors should be located at least 30 mil from the edge of the substrate.
8. High-power resistors should be evenly distributed on the substrate and should not be located near the substrate edge or close to active components.
9. Resistors with the same sheet resistivity and similar geometries will track much better with temperature than resistors of different resistivity or geometry.
10. These constraints are only guidelines for optimum design and are not hard and fast rules. In some applications they

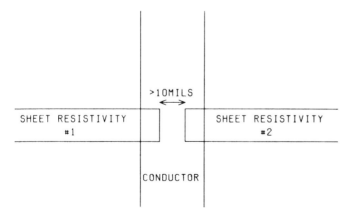

FIGURE 4.9. Termination of different sheet resistivities.

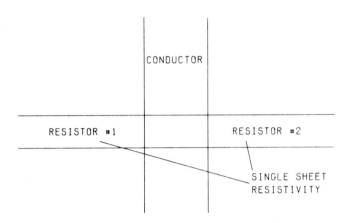

FIGURE 4.10. Termination of resistors with the same sheet resistivity on a common pad.

are violated for many reasons. However, in these cases it is important to remember that other effects may come into play and the process tolerances may be much worse than ±20%. Most of these effects are material dependent and are difficult to predict and characterize.

Typical Characteristics of Thick-Film Resistors

1. *Temperature coefficient of resistance (TCR)*: The best resistor films have TCRs of ±100 ppm/°C over -55 to 125°C.
2. *TCR tracking*: Resistors of the same paste in the same screening step will track to ±15 ppm/°C.
3. *Long-term stability*: Well-designed resistors will drift less than ±0.25% after 5000 hr at 125°C. The tracking between resistors will be better than 0.1% after 5000 hr at 125°C.
4. *Power dissipation*: Most commercial resistor pastes are specified by the manufacturers as capable of dissipating power densities of up to 50 W/in.^{2}. However, this assumes that the total power dissipation on the substrate does not exceed 1 to 2 W/in.^{2}. As will be discussed in Chapter 10, this figure is slightly misleading and may be greatly exceeded in some designs, particularly for minimum geometry resistors. It is recommended that a detailed thermal analysis is made for all but the simplest hybrid designs.

Thick-Film Design Guidelines

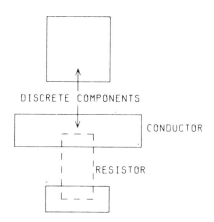

A) POOR ATTACHMENT LAYOUT

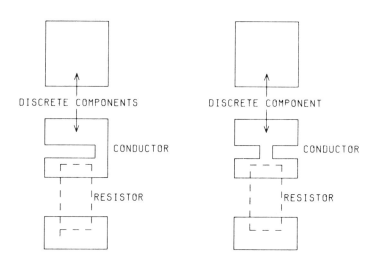

B) GOOD ATTACHMENT LAYOUTS

FIGURE 4.11. Terminations for resistors and discrete components sharing a common pad.

4.4. BASIC DESIGN RULES FOR THICK-FILM CONDUCTORS

Description of Thick-Film Conductors

Conductors usually constitute the largest quantity of material screened on a thick-film circuit. There are many conductor compositions each with different properties. Various functions needed are solderable connections, resistor terminations, crossover connections, capacitance electrodes, wire bond pads, and eutectic die bond pads.

Particular materials are available for each of these functions and should be used accordingly. Materials in general usage are gold for good conductivity and bonding pads, platinum gold for optimum solder leach resistance, and palladium silver for low-cost and solderable conductors. Table 4.1 summarizes the properties of the most commonly used thick-film conductors.

General Conductor Design Guidelines

Conductors can be as narrow as 8-mil lines and 7-mil spaces for thick film. Conductors with 5-mil lines and 5-mil spaces can be fabricated with some yield loss. Where possible lines should be oriented in the X-Y direction, although diagonal lines and curved lines can be fabricated. Line widths should be as wide as possible and as short as possible if resistance could be a problem. Conductor areas should be minimized where parasitic capacitance could be a problem.

Conductors can be screened on both the front side and the back side of the substrate, and can be connected through holes in the substrate or around the edges. However, connection of the conductors front to back is difficult and results in yield losses and potential failure mechanisms.

TABLE 4.1. Properties of Thick-Film Conductors

Material	Cost	Sheet R $m\Omega$ per square	Solderable	Wire bondable	High-frequency performance
Gold	High	3-4	No	Good	Good
Platinum gold	Medium	50-80	Yes	Poor	Poor
Palladium silver	Low	25-35	Yes	Poor	Poor

Thick-Film Design Guidelines

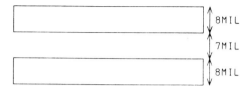

FIGURE 4.12. Minimum conductor line widths and spacings.

Margin Considerations

In the printing operation there will be a part-to-part alignment error of about ±5 mil caused by the inability to always position a part exactly the same in the printer. Closer alignment is possible but will add cost to the part due to the additional time required. To compensate for this variation a margin or border must be left around the outside of the pattern. The necessary margin criteria are shown in Figure 4.3.

Conductor Lines and Spaces

Line widths and spacings should be as large as practical. Material costs, run resistance, and stray capacitance must be considered when designing conductor runs. The wider the runs, the easier the part is to fabricate, the larger the capacitance and material cost, and the lower the run resistance, and vice versa.

The minimum line widths and spaces that can be easily fabricated are 8-mil line widths and 7-mil spaces, as shown in Figure 4.12.

Lower line widths and spaces are possible but lower yield due to opens and shorts will increase the cost of the hybrid. The present minimums are about 5-mil-wide lines and 5-mil spaces.

All conductor lines should be designed as straight lines, not curved, since it is usually difficult to generate curves on the artwork. Curved lines are used in some cases but only if there is a compelling reason due to circuit parameters.

Description of Output Lead Pads

External leads are usually soldered to the substrate, and the conductor pads for attachment should be as large as possible to provide adequate strength. Many commercial leads come in long lengths on 0.100-in. centers. Typical pad dimensions for these leads are shown in Figure 4.13. Pads are usually screened on both sides of the substrate to give maximum strength.

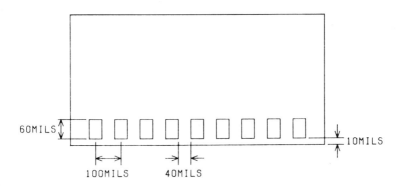

FIGURE 4.13. External lead pad dimensions.

Description of Conductor-to-Conductor Interfaces

When two conductor systems are printed in different screenings, a possible registration error can occur. This is usually of the order of ±5 mil. Figure 4.14 shows examples of poor and correct designs of overlapping conductors. The correct design includes a 5-mil margin to allow for registration error.

Conductor Pads Around Holes in the Substrate

Holes are often used in substrates for mounting of discrete components or for making connections from one side to the other. Solder pads surrounding holes should be square where possible, since this facilitates screen fabrication and processing, although round pads are also acceptable. The pad length and width (or diameter) should be at least three times the diameter of the component to be soldered with a 10-mil minimum width of conductor. A margin of at least 5 mil, and preferably 10 mil, should be left around the hole to avoid blocking the hole with paste during the screening operation. Figure 4.15 shows typical dimensions of hole/screen configurations.

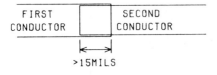
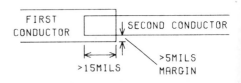

FIGURE 4.14. Overlapping conductor geometries.

Thick Film Design Guidelines

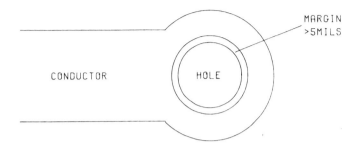

FIGURE 4.15. Conductor terminations around holes.

Die and Wire Bond Pads

Wire bonds can be made to any gold conductor that is 10 mil wide or wider.

The die pad on which an unpackaged integrated circuit chip is to be bonded must be at least 20 mil larger than the chip that is to be bonded (10 mil allowed on each side). It is important that prior to layout, the chip manufacturer's catalog be consulted for maximum possible chip size. Figure 4.16 shows typical die attach wire bond configurations.

Reflow-Soldered Packaged Device Pads

In a solder reflow process, individual discrete components are aligned with prewetted conductor pads and then placed in a special furnace where the solder remelts, fixing the device in position. Proper con-

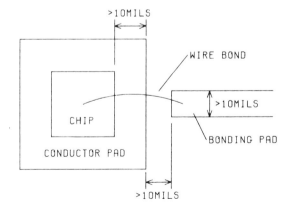

FIGURE 4.16. Die and wire bond pads.

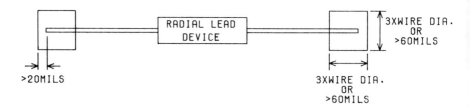

FIGURE 4.17. Attachment or radial lead device.

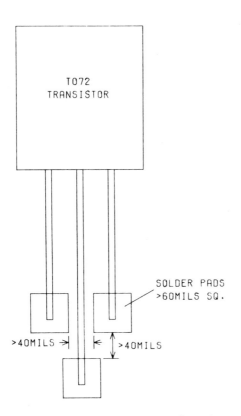

FIGURE 4.18. Attachment of surface-mounted transistors.

ductor and encapsulation placement helps the assembler align the components and minimizes the yield loss due to solder bridges.

The width of pads should be at least three times the lead wire diameter to be attached. The pad length is usually determined by the type of component being attached.

Examples of typical attachment methods for various components are shown in Figures 4.17 to 4.23.

Good design practice calls for 5-mil encapsulation relief around device pads, and spacing of at least 15 mil from solder pads to the nearest conductor.

Edge-mounted transistors as shown in Figure 4.19 can serve as a dual purpose. They give the device some mechanical strength and can often serve as a convenient interconnection between front and back circuitry on the substrate.

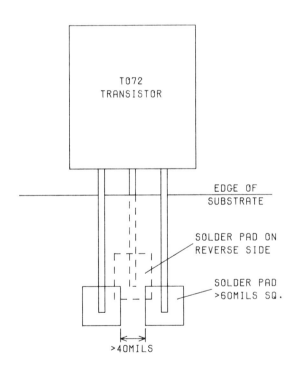

FIGURE 4.19. Attachment of edge-mounted transistors.

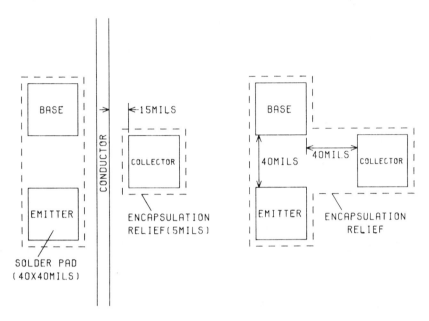

FIGURE 4.20. Attachment of SOT-23 transistors.

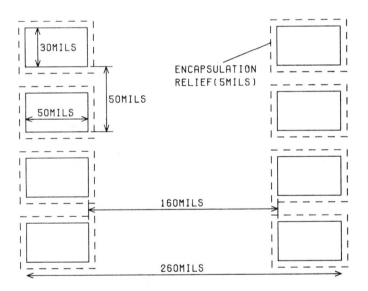

FIGURE 4.21. Attachment of SOT 8-pin IC packages.

Thick-Film Design Guidelines　　55

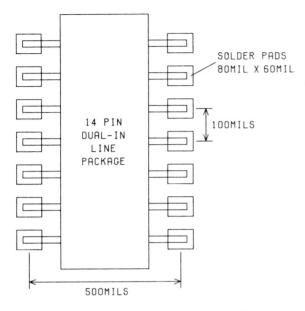

FIGURE 4.22. Attachment of dual-in-line packages.

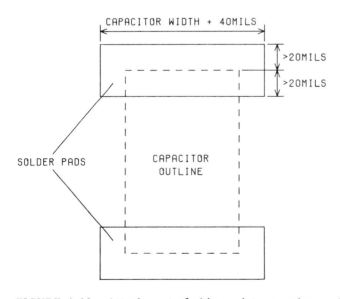

FIGURE 4.23. Attachment of chip resistors and capacitors.

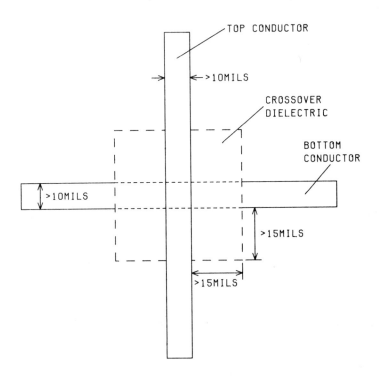

FIGURE 4.24. Thick-film crossover.

Epoxy-Mounted Device Pads

Occasionally, reflow soldering may not be possible and components can be attached using conductive epoxies. Pad dimensions are usually similar to those for soldered components.

Thick-Film Crossovers

One advantage of the thick-film process is its ability to provide reliable and inexpensive crossovers with high yield. A crossover is made by printing a conductor, followed by an insulating (dielectric) layer, followed by a second "crossover" conductor. The structure is shown in Figure 4.24.

The process involved is:

1. Screen first conductor, dry
2. Fire first conductor
3. Screen dielectric layer, dry

Thick-Film Design Guidelines

4. Screen second conductor, dry
5. Co-fire dielectric and second conductor

A consideration with such a crossover is that capacitance between the first and second conductors could cause cross-coupling in a high-frequency circuit. A typical dielectric constant of 9.4 corresponds to a capacitance of approximately 0.0015 pF/mil^2 for 1.4-mil fired dielectric thickness. For example: For a 10-mil conductor crossing a 10-mil conductor the coupling capacitance will be approximately 0.15 pF. The breakdown voltage for a crossover will usually exceed 500 V.

Typical layout criteria for thick-film crossovers are:

1. First conductor not less than 10 mil.
2. Dielectric must extend at least 15 mil beyond either conductor to prevent shorting.
3. One conductor *cannot* be palladium silver if the other is gold or platinum gold.
4. Keep top conductor as short as possible to obtain high yields.

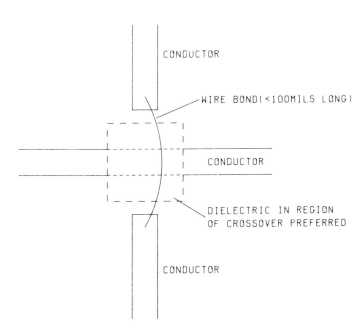

FIGURE 4.25. Wire-conductor crossovers.

Wire Conductor Crossovers

If only one or two crossovers are required on a substrate, it may be more economical to design for wire-bonded crossovers than to involve the extra screenings for a thick-film crossover. When wire crossovers are used, a dielectric encapsulation is usually placed underneath the bond to insulate the wire bond from the conductor being crossed. Figure 4.25 shows typical dimensions. Wire bonds should not exceed about 0.100 in. in length.

4.5 BASIC DESIGN RULES FOR THICK-FILM CAPACITORS

Description of Thick-Film Capacitors

Thick-film capacitors can be made in several ways, values from 0.1 to 10,000 pF being feasible. Dielectric pastes are available with dielectric constants ranging from 10 to 20,000. The low-K dielectrics are very stable but the high-K dielectrics have poor temperature coefficients and stability.

Substrate Capacitors

A convenient way of fabricating low values of capacitance is to utilize the substrate itself as the dielectric, as shown in Figure 4.26.

The capacitor is to ground and its value is a function of the substrate dielectric constant and thickness:

$$\text{capacitance } C = 0.225 \frac{\varepsilon_r A}{t}$$

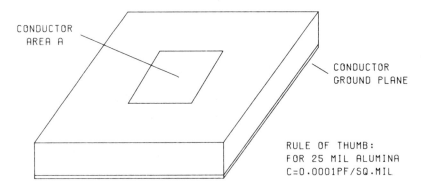

FIGURE 4.26. Substrate capacitor.

Thick-Film Design Guidelines

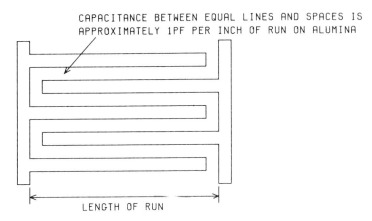

FIGURE 4.27. Interdigital thick-film capacitor.

where

C = capacitance, pF
ε_r = dielectric constant
A = plate area, in.^{2}
t = thickness of substrate, in.

For alumina dielectric with a dielectric constant of 9.6 and 25 mil thick, the capacitance is 8.6×10^{-5} pF/mil^2. (A good rule of thumb is 10^{-4} pF/mil^2 for the standard 25-mil alumina substrate).

This type of capacitor is good for values in the region 1 to 3 pF. Above these values the capacitors take up too much substrate area. For example, a 1 pF capacitor is approximately 100×100 mil.

Interdigital Capacitors

This configuration is shown in Figure 4.27. The capacitance is approximately 1 pF per inch of run for an alumina dielectric with equal line widths and spaces. Typical values are from 1 to 3 pF. Above this value the capacitors take up too much area. A major limitation is the minimum line width and spaces of 5 mil generally used. This type of capacitor is better suited to the fine-line definition of thin films.

Parallel Plate Capacitors

The configuration shown in Figure 4.28 consists of a first conductor, dielectric layer, and a second conductor. The capacitance is given by

$$C = 0.225 \frac{\varepsilon_r A}{t}$$

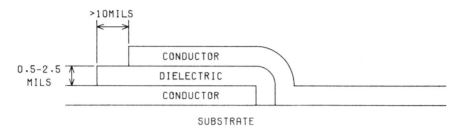

FIGURE 4.28. Parallel plate thick-film capacitor.

where

C = capacitance, pF
ε_r = dielectric constant of dielectric
A = area, in.^{2}
t = dielectric thickness, in.

The usual dielectric thickness t is from 0.5 to 2.5 mil, a nominal of 1.4 mil being usual.

The dielectric should overlap the top and bottom electrodes by 10 mil on all edges. With a dielectric constant of 100 and with 1.4-mil thickness, a 50 × 50 mil capacitor will have approximately 40 pF capacitance.

Capacitor Tolerances

The substrate-to-ground plane capacitor will have about a 10% tolerance due to variances in substrate thickness and dielectric constant. The interdigital capacitor tolerance is a function of processing and dielectric constants and can be of the order of 5 to 10% with precision processing. The parallel plate thick-film dielectric capacitor will have a tolerance of about ±30% due to wide variations in fired dielectric thickness.

Typical Thick-Film Dielectric Capacitor Characteristics

Typical characteristics of capacitors using thick-film dielectrics are:

Quality factor Q at 1 MHz	1000 to 1650
Dissipation factor	0.05 to 0.25%
Insulation resistance	>10^{12} ohms (100 V dc)
Breakdown voltage	>400 V ac rms
Temperature coefficient of capacitance	<200 ppm/°C from −55 to 125°C

Thick-Film Design Guidelines

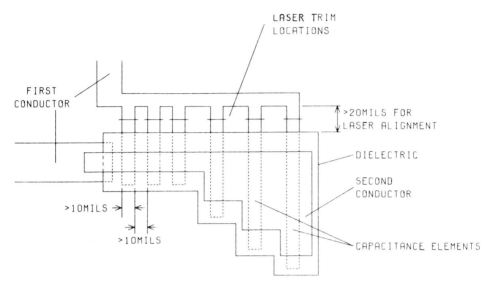

FIGURE 4.29. Laser-trimmed thick-film capacitor.

Variable Thick-Film Capacitors (Digitally Trimmed with Laser)

A frequent requirement in thick-film high-frequency circuits is for a variable capacitor. Laser trimming provides a method of removing area from a thick-film capacitor, thereby reducing its value. A useful configuration for this purpose is shown in Figure 4.29, where the capacitor can be reduced in discrete steps. Analog trimming of the parallel plate thick-film capacitor is not recommended due to the high probability of shorts between the top and bottom electrodes across the laser kerf. The capacitor shown in Figure 4.29 is made up of several parallel capacitors, and the capacitor is trimmed down in value by cutting out selected loops.

Because capacitors are trimmed down in value, they are designed higher in value than the desired capacitance by an amount determined by the estimated processing variation. This ensures that the as-processed capacitor will be greater than or equal to the required capacitance. Thus if the final capacitance required is C_f, and the process tolerance is $\pm P_{tol}$, the capacitor would be designed in layout to be $C_f/(1 - P_{tol})$.

5

THIN-FILM MATERIALS

5.1 INTRODUCTION

Thin-film materials are deposited or sputtered on a ceramic or insulating substrate. The films can be conductive, resistive, or insulating and are generally less than 1 μm thick, compared with the 10 to 50 μm of thick-film materials. Thus thin-film technology can be used to produce resistors, capacitors, and conductors on a substrate. It is also possible to produce thin-film active devices, although little commercial success has been achieved in this area. The advantages of thin-film technology are that better resolution and higher densities can be achieved. The high resolution of thin film has led to extensive use of this technology in microwave integrated circuitry, where high resolution and low losses are required on transmission lines and other microwave elements. Another major area of application is in precision resistor networks for A/D and D/A converters.

5.2 THIN-FILM PROCESSES

The most common processing techniques for producing thin-film circuits are vacuum evaporation, cathode sputtering, vapor-phase deposition, and plating techniques. Typical vacuum evaporation and cathode sputtering systems are shown in Figure 5.1.

Vacuum Evaporation

Vacuum deposition is usually carried out in a bell jar under high-vacuum conditions, usually 10^{-5} to 10^{-6} torr.

The vacuum systems are often diffusion pumped and have liquid nitrogen cold traps.

The substrate and material to be evaporated are placed in the bell jar, the system pumped down, and the material heated by an electrical

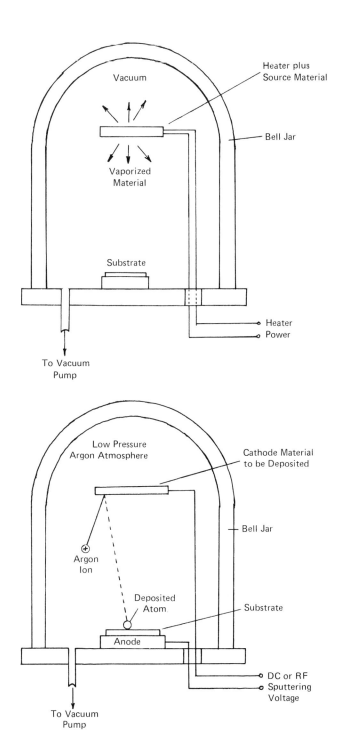

FIGURE 5.1. Vacuum evaporation and cathode sputtering systems.

element until it vaporizes. Under the high vacuum, the mean-free path of the vaporized molecule is comparable to the dimensions of the bell jar and the vaporized material radiates in all directions within the bell jar. The substrate is placed some distance from the vaporized source and some of the vaporized material deposits on it to form a fairly uniform thin film. The substrate is usually heated to provide good adhesion of the deposited film.

The vacuum evaporated films exhibit a fine-grained structure which is improved at high evaporation rates. The grain structure, uniformity, and repeatability of the deposited film are improved if the angle of incidence of the radiating vapor on the substrate is made steeper.

Both resistive and conductive films can be deposited by vacuum evaporation, including aluminum, gold, and silver conductors and nickel-chromium resistors.

In cases where high energies are required to vaporize the source material, electron beam bombardment is used rather than thermal heating. The electron beam evaporation technique allows deposition of most conductors, dielectrics, and resistor materials. However, alloy depositions such as nickel-chromium resistors are difficult to control because each alloy component has a different evaporation rate at a given temperature.

Cathode Sputtering Using a Glow Discharge (DC Sputtering)

Cathode sputtering takes place in a low-pressure, inert gas atmosphere, usually argon. A glow discharge is formed by applying a high dc voltage in the range 2000 to 5000 V across the cathode and anode. Due to the dc voltage the process is often called dc sputtering. The cathode is the source of sputtered atoms and constitutes the material to be deposited. The substrate is attached to the anode or placed within the glow discharge region. The positively charged argon ions are accelerated toward the cathode due to its negative potential. These ions bombard the cathode material with high energy and cause the atoms or molecules of the cathode material to break away, or sputter, from the cathode. Some of these molecules are intercepted by the substrate and form a very uniform thin-film layer. The mean free path of the source atoms is much shorter than in vacuum evaporation due to the relatively higher gas pressure of the argon gas. This results in lower deposition rates for the sputtering process compared with vacuum evaporation.

The chemical composition of the deposited films can be modified by adding small amounts of reactive gases such as oxygen or nitrogen to the inert argon atmosphere. The properties of the deposited films can be altered, depending on the concentration of the added reactive gases. This process is known as *reactive sputtering*.

Thin-Film Materials

Radio-Frequency Sputtering (RF Sputtering)

Insulators cannot be sputtered through conventional dc sputtering techniques because the accelerating potential cannot be directly applied to the insulator surface. This prevents neutralizing of the positive charge which would accumulate on the surface during ion bombardment. This problem can be overcome by applying a high-frequency potential to a metal electrode behind the insulator. Power is fed to the plasma by capacitive coupling through the insulator dielectric.

Sputtering can occur because the insulator will now be alternatively ion bombarded and electron bombarded. The positive charge which accumulates on the surface during the negative or sputter portion of each cycle will now be neutralized by electrons during the positive half of the cycle. Because of high frequencies normally employed, the process is called radio-frequency (RF) sputtering. The RF frequency is usually at the industrial, scientific, and medical (ISM) frequency of 13.56 MHz, since this frequency has less critical specifications with regard to RF radiation than most others.

RF sputtering is also frequently used for depositing resistive films such as nickel-chromium. A major advantage of sputtering over vacuum evaporation is that the sputtered films can retain the target stoichiometry. This is particularly important in the deposition of alloy films such as nickel-chromium.

RF sputtering has advantages over dc sputtering in that the target material can be a dielectric, resistor, or conductor, and the lower pressures usually employed in RF sputtering result in less entrapment of the sputtering gas in the deposited films.

Sputtered films also show better substrate adhesion properties than evaporated films. The limiting factors on materials that can be sputtered are the vacuum compatability of the material and the availability of commercial targets.

Vapor-Phase Deposition

In vapor-phase deposition halide compounds of the materials to be deposited are chemically reduced to form the required metal deposition on the substrate. The process is very similar to the epitaxial process in semiconductor processing.

The process is usually used to deposit thick layers of material, up to 20 μm. Aluminum oxide dielectric and tin oxide (SnO_2) resistive films are commonly deposited using vapor-phase deposition.

Plating Techniques

There are two types of plating techniques used in thin-film processing, electroplating, and electroless plating. In electroplating, the substrate is made the cathode terminal and is immersed in an electrolytic plating solution. An electrode made up of the metal to be plated serves as the anode. When a direct current is passed through the solution, positively charged metal ions migrate from the anode and deposit at the cathode. The required conductor pattern is usually formed by electroplating the metal through a precision photoresist image onto previously vacuum deposited metal films.

In electroless plating simultaneous reduction and oxidation of a chemical agent is used in forming a free metal atom or molecule. Since this method does not require electrical conduction during the plating process, it can be used directly with insulating substrates without previously vacuum depositing a metal film.

Electroplating is commonly used to form films of gold or copper, and electroless plating is used for plating nickel, copper, and gold.

In most commercial thin-film hybrid applications electroplating is usually used to increase the thickness and the conductivity of previously vacuum deposited metal films.

General Comments on Thin-Film Deposition Techniques

While it would seem that a thin-film of almost anything can be deposited on everything, there are limitations:

1. The film must adhere to the substrate and not react with it to the point that the desired properties of either are altered.
2. The film or films must be compatible with photoprocessing.
3. The film or films must withstand any process temperatures they may encounter. Film adhesion should not be lost nor undesirable alloys formed at temperatures encountered during die attach, wire bonding, and other postdeposition processes.
4. The films must withstand the environment they will be used in.

5.3 THIN-FILM CONDUCTOR MATERIALS

The most common thin-film metal for conductors is gold. Usually, gold is vacuum deposited to a thickness of about 200 to 500 nm. The sheet resistivity of this thin-film gold is about 0.025 to 0.050 Ω per square. If the sheet resistivity is too high, the conductor areas are electroplated through a photoresistive image to about 10 μm thick. This gives a sheet resistivity of about 0.004 Ω per square.

Other materials that have been used for thin-film hybrids are silver, aluminum, and tantalum. These materials have definite cost advantages over gold but have compatability problems with other hybrid processes such as die attach and wire bonding.

5.4 THIN-FILM RESISTOR MATERIALS

The most common thin-film resistor systems are nickel-chromium (nichrome) and tantalum nitride. These materials have low sheet resistivity, usually less than 200 Ω per square. Tin oxide (SnO_2) is often used where higher sheet resistivities from 100 to 5000 Ω per square are required.

RF sputtering of nickel-chromium films is preferred to vacuum deposition since the RF sputtered film retains the composition ratio of the source material. The ratio of nickel to chromium affects the resistivity and temperature coefficient of resistance. Composition ratios of 60:40 nickel to chromium give low TCRs in the region 50 to 100 ppm/°C. Although sheet resistivities up to about 200 Ω/square can be fabricated, the upper limit for good control is about 75 to 100 Ω per square. Tantalum nitride films have proven to have very reproducible properties for thin-film resistors. Tantalum is sputtered in the presence of nitrogen and this results in reactively sputtered films of tantalum nitrides (Ta_2N). Typical reactive gas concentrations are in the range 0.5 to 5% nitrogen. Tantalum nitride films are usually in the sheet resistivity range 25 to 100 Ω per square, with negative TCRs less than 100 ppm/°C.

Tin oxide (SnO_2) films are usually deposited using vapor-phase deposition. The sheet resistivity of these films can be controlled by group III or group V ions such as indium (In) or antimony (Sb) to increase or reduce the sheet resistivity.

5.5 THIN-FILM DIELECTRIC MATERIALS

Dielectric films of silicon monoxide, silicon dioxide, and tantalum pentoxide are commonly used as thin-film dielectrics. Silicon monoxide films are usually vacuum evaporated and have a dielectric constant of about 5. Silicon dioxide can be sputtered and makes a fairly good dielectric film with dielectric constant of about 4.

Resistive films of tantalum can be anodized to form a dielectric film of tantalum pentoxide with a dielectric constant of about 25. A major advantage of a tantalum-based thin-film system is that both resistive and dielectric films can be easily obtained.

5.6 THIN-FILM PROCESSING (Conductor-Resistor Networks)

Hybrid thin-film processing consists of a series of process steps involving deposition, etching, and plating. The two processes most commonly used are pattern plating and thin-film etchback.

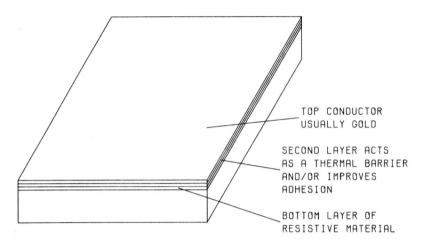

FIGURE 5.2. Thin-film metallized substrate.

Pattern Plating

Thin-film resistor material is first deposited on the substrate, followed by deposition of the metal conductor layer. If the conductor is gold, often intermediate layers are deposited to improve the adhesion to the resistor material. Typical intermediate layers are titanium, palladium, and nickel. The process is shown in Figure 5.2. Photoresist is then applied to the substrate by spin or dip coating and the required conductor layer is patterned using photolithographic techniques. The thin-film gold is then electroplated to form the required conductor pattern. The gold is normally plated to about 0.5 mil thickness. The photoresist, thin-film gold, and intermediate layers are chemically removed, leaving electroplated conductors on the resistor layer. Figure 5.3 shows the three steps in this process. The final step is to reapply photoresist and pattern using photolithographic techniques to protect the resistor areas. The exposed resistor material is then etched away and the photoresist chemically removed. The final circuit, including plated conductor and resistor, is shown in Figure 5.3.

Thin-Film Etchback

If ultra-precision line widths and resistor values are required, the thin-film etchback process is used. Some sacrifice is made in conductor resistivity since this process does not use plated conductors. The gold metal is deposited to about 1/2 μm thickness, which is considerably less than the 8 to 10 μm of electroplated gold in the pattern plating process.

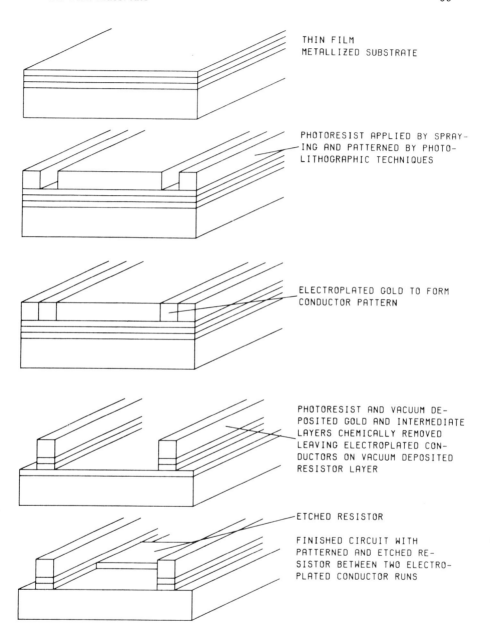

FIGURE 5.3. Thin-film electroplated conductors.

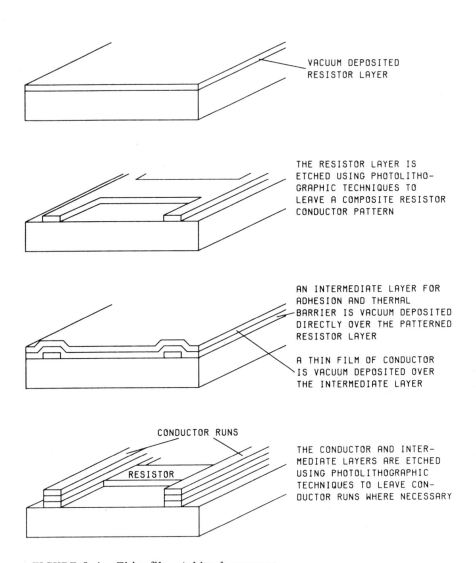

FIGURE 5.4. Thin-film etchback process.

Thin-Film Materials 71

The process consists of depositing the resistor layers and defining the composite resistor-conductor pattern using photolithographic techniques and etching. The thin metal layers are then deposited and the conductor patterns defined. The process is shown in Figure 5.4.

5.7 ENVIRONMENTAL PROTECTION OF THIN-FILM RESISTORS

The completed thin-film hybrids require some form of environmental protection. The final assemblies can be protected by hermetically sealing within a ceramic or metal package. However, such sealing is often difficult and expensive. Other less expensive methods are available to protect the thin-film elements from the environment.

Tantalum nitride resistors can be protected by anodizing the resistors to build up an oxide surface. The resistor material is deposited at a lower sheet resistivity than the design value. The resistor material is then thermally stabilized and finally anodized to the design value. The anodization process can be mechanized with feedback control to terminate the process when the desired value is reached. Resistors can be anodized to about 0.1% accuracy. The anodized surface provides a high resistance to electrochemical erosion under humid conditions and dc potential. Tests of such resistors at 100% humidity and 65°C operating temperature show drifts of less than 0.05% after 2000 hr. This compares very favorably with the best discrete resistors under the same conditions.

Polyimide films can also be used to protect thin-film circuits, and are particularly useful for encapsulating Nichrome resistors. The material in the form of a polyamic acid solution is spun on the substrate and baked at low temperature. The film can then be patterned using photolithographic techniques, and fired at 250°C to form polyimide film. The film is used to protect Nichrome resistors and fine-line conductors, and has also been used to provide solder dams. Nichrome resistors protected with polyimide film show similar stabilities to the tantalum nitride films.

6
THIN-FILM DESIGN GUIDELINES

6.1 DESCRIPTION OF THIN-FILM RESISTORS

The ideas presented in Section 4.3 also apply to thin-film resistors. The value of a resistor is determined by the sheet resistivity and the length-to-width ratio (aspect ratio). Usually, the substrate is coated completely with resistive material and then conductive material. The conductors and resistors are then etched back using appropriate photoetching techniques. Because of the accuracy obtained in photoprocessing, very fine line widths can easily be achieved. Typical line widths can be as low as 1 to 2 mil. The film deposition and uniformity can be controlled to about ±10%, giving as processed resistors with tolerances of about 10 to 15%. Most thin-film resistors can be brought into about ±5% tolerances with heating at elevated temperatures; this is caused by annealing or by oxidation of the surface metal. This heating at elevated temperatures also helps produce very stable resistors.

One disadvantage of thin-film resistors compared with thick film is that resistive thin films have a limited range of sheet resistivities. Tantalum nitride and Nichrome have low sheet resistivities (25 to 200 Ω/square). Materials such as chrome-silicon monoxide have been used for higher sheet resistivities of about 1 to 10k Ω per square.

Resistor Design Guidelines

Usually, with thin-film technology the designer has only one sheet resistivity available. This can place some restrictions on the design. For example: if a 50-Ω per square film is available, it would take 20,000 squares to design a 1-MΩ resistor. Even at 1-mil line widths, this would mean a meander resistor with total length of 20 in., not a very feasible design. Thus for hybrids with very high ratios of maximum to minimum resistor values, thick film is usually the better technology.

However, with thin films, 1000 squares would not be unreasonable to achieve. Thus with a 50-Ω per square system, resistors from 10 Ω to 50 kΩ can be achieved fairly easily.

Other restrictions on resistor size may be power dissipation and voltage stress. Power density should be less than 50 W/in.^{2} and voltage stress less than 1 V/mil. Under certain circumstances power densities much greater than 50 W/in.^{2} can be used. Power density considerations are discussed in greater depth in Chapter 10.

Resistors should be designed as wide as possible to improve yields, although resistors down to 1 mil wide with 1 mil of space can be achieved with reasonable yields.

Resistor Tolerances

Due to variations in deposition thickness and photoprocessing tolerances, as processed resistors can be obtained to about ±10%. Closer tolerances to about ±5% can be achieved by designing the resistor about 10% on the low size and annealing or oxidizing the resistor into tolerance. Closer tolerances down to ±0.1% can be achieved with laser trimming. If laser-trimmed resistors are required, they are designed 15 to 20% low in the layout stage, with about 40% trim capability on the resistor.

Typical resistor trim geometries are shown in Figure 6.1.

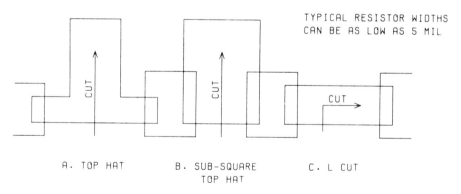

FIGURE 6.1. Thin-film resistor geometries.

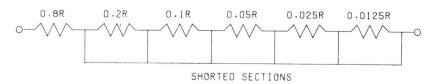

FIGURE 6.2. Schematic of a digitally trimmed thin-film resistor.

Typical Thin-Film Resistor Characteristics

For nichrome resistors:

Temperature coefficient of resistance (TCR)	+60 ppm
TCR tracking	± 5 ppm
Long-term stability	0.01%

For tantalum nitride resistors:

Temperature coefficient of resistance	-75 ppm
TCR tracking	± 5 ppm
Long-term stability	0.01%

Close-Tolerance Thin-Film Resistors

With analog trimming of top hats, thin-film resistors with 0.1% tolerances can be achieved at the end of life. A major contributor to the long-term drift is the exposed resistor at the laser trim cut. Tighter tolerance resistors can be achieved using digital trim techniques. This is achieved by adding shorted resistor sections to the main meander resistor. Each resistor section is a certain percentage of the total resistor value. If a shorted section is opened with the laser, the resistor will increase without exposing any new resistor material. A typical example is shown in Figure 6.2.

Using a computer-controlled laser, the resistor is initially measured and the appropriate resistor sections added sequentially by opening up their shorting bars. Because of process tolerance variations some redundant sections usually need to be built in. Figure 6.3 shows a typical layout for a precision resistor.

Using this discrete trim approach resistors of very high stability with end-of-life tolerances of ±0.01% have been obtained.

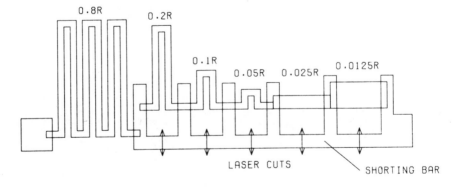

FIGURE 6.3. Layout of a digitally trimmed thin-film resistor.

6.2 THIN-FILM CONDUCTOR GUIDELINES

The most usual thin-film conductor material is gold. The conductor pattern is usually photoetched using photolithographic techniques. Two methods are generally used, depending on the conductor resistivity constraints in the design.

Thin-film gold usually has a relatively high sheet resistivity of 0.025 to 0.050 Ω per square. If this is not a problem, the thin-film conductor is photoetched in the required pattern. If this sheet resistivity is too high, the conductor areas are initially plated to 0.3 to 0.5 mil thickness (comparable to thick film). The remaining thin-film gold is then etched away. The plated-up gold has sheet resistivities from 0.003 to 0.005 Ω per square.

Thin-film conductors can be defined with about 0.1-mil edge variation, and can be designed down to 2-mil widths with 2-mil gaps when plated up. In special applications, nonplated thin-film conductors can be designed down to a few micrometers. The constraints on pad dimensions for components are approximately the same as for thick-film conductors.

Through-Hole Connections

During deposition of the gold conductor it is possible to make a reliable metal connection from front-to-back metallizations on the substrate. Electroplating can reduce the resistance of through-hole connections to a few milliohms.

Typical hole dimensions are 20 to 25 mil for a 25-mil alumina substrate. The hole edges need to be fairly well defined to obtain reliable connection. Good yields can be obtained with both diamond-drilled and laser-machined holes. Through-hole connections are particularly useful in high-frequency hybrids where connections to the back metallization, usually a ground plane, need to be made as close as possible to the device to be grounded.

6.3 DESCRIPTION OF THIN-FILM CAPACITORS AND CROSSOVERS

Alternative Configurations for Thin-Film Capacitors

Thin-film capacitors and crossovers can be made in several ways. Generally, high-dielectric-constant materials are not available and only low values from 0.1 to 1000 pF are obtained.

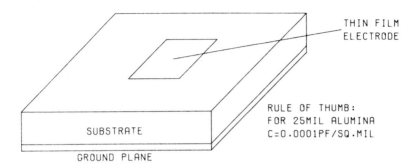

FIGURE 6.4. Thin-film substrate capacitor.

The substrate can be used to obtain small values of capacitance from the front surface to the back surface of the ground plane. This is shown in Figure 6.4, the capacitance is given by

$$C = 0.225 \frac{\varepsilon_r A}{t}$$

where

C = capacitance value pF
ε_r = dielectric constant
A = plate area, in.^{2}
t = thickness of the substrate, in.

Capacitors can be obtained in the range 0.1 to 3 pF with approximately 20% tolerances.

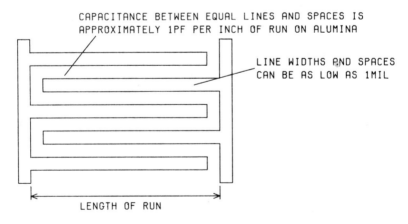

FIGURE 6.5. Thin-film interdigital capacitor.

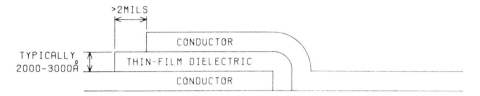

FIGURE 6.6. Thin-film parallel plate capacitor.

Higher-value capacitors can be obtained using the interdigital structure shown in Figure 6.5. Thin film can provide higher values per unit area with the interdigital capacitor than thick film, due to the finer line widths obtained with thin film. The thin-film structure can be made with conductors down to about 1-mil line width and 1-mil spacing. The capacitance for alumina substrates is about 1 pF/in. of conductor run, and values to about 5 pF can be obtained, with about 20% tolerances.

The parallel-plate-capacitor approach can be used for both capacitors and crossovers. Figure 6.6 shows the deposition layers involved: the lower conductor area, the dielectric film, and the upper plate areas. The dielectric film is very thin compared with thick-film dielectrics, and this compensates somewhat for the relatively low-dielectric-constant materials used. Values have been obtained in the range 1 to 100 pF.

Dielectrics that have been used for thin-film capacitors are silicon monoxide ($\varepsilon_r = 6$), aluminum oxide ($\varepsilon_r = 10$), and tantalum pentoxide ($\varepsilon_r = 26$). Silicon monoxide and aluminum oxide are usually sputtered, whereas tantalum pentoxide is usually formed by thermal or anodic oxidation of a tantalum film. The original tantalum film forms the bottom electrode. Because of the thin layers of dielectric the breakdown voltages are usually low and the yield loss due to pinhole shorts is also a problem. Often, a dual dielectric consisting of silicon monoxide and one of the other dielectrics is used to reduce yield losses and

FIGURE 6.7. Dual dielectric thin-film capacitor (tantalum pentoxide/silicon monoxide).

increase voltage breakdown. However, this will obviously result in lower capacitance per unit area.

A viable tantalum pentoxide/silicon monoxide capacitor is shown in Figure 6.7. Tantalum is first deposited to form the base electrode. This is then anodized to form about 10 μm of tantalum pentoxide. About 200 μm of silicon monoxide is then deposited to form the second dielectric. The final top electrode is formed by depositing gold. The resulting capacitor has about 0.17 pF/mil^2; a 100 × 100 mil capacitor has about 1700 pF with about 25% tolerance. Typical breakdown voltage is about 20 V.

Typical Thin-Film Capacitor Characteristics

Quality factor Q at 1 MHz	500 to 2000
Dissipation factor	0.1 to 0.25%
Insulation resistance	>10^{10} Ω
Breakdown voltages	>20 V
Temperature coefficient of capacitance	200 ppm/°C

One alternative to using thin-film materials for capacitors is to use the polyimide discussed in Chapter 5. Gold can be used as the bottom conductor and polyimide film spun on and processed using photolithographic techniques. The film is then baked and the top electrode deposited. The final capacitor is shown in Figure 6.8. Because of the low dielectric constant of the polyimide (3.4) and the relatively thick dielectric (typically about 5 μm), the capacitor has a low capacitance per unit area, approximately 3.2$_{10}$ - 3 pF/mil^2. Thus a 100 × 100 mil capacitor will be approximately 32 pF with about ±30% tolerance.

Laser Trimming of Thin-Film Capacitors

Thin-film capacitors can be processed as multiple parallel capacitors as described for thick films. The capacitance is reduced by laser trimming and removing discrete sections. A typical configuration for laser trimming is shown in Figure 6.9.

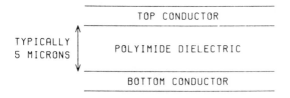

FIGURE 6.8. Polyimide thin-film capacitor.

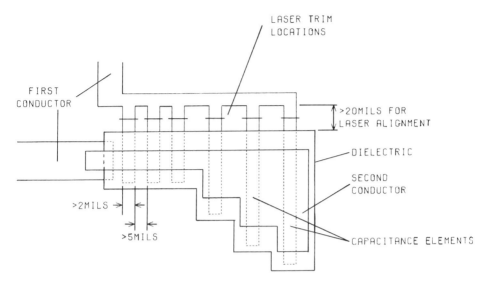

FIGURE 6.9. Laser-trimmed thin-film capacitor.

It is not advisable to trim the bulk capacitor because of the high probability of shorts and associated reliability problems. An alternative method for trimming very low value capacitors is to use the interdigital structure as shown in Figure 6.5 and to reduce its value by open-circuiting digits with the laser. Precision trimming of the frequency response of high-frequency circuits can be achieved with this technique.

6.4 THIN-FILM SNAPSTRATES

Thin film is an inherently high-cost process when compared with thick film. However, because of the high density possible with thin film, it may be possible to reduce the hybrid size significantly for the same circuit configuration. For example: a 1 × 1 in. thick-film circuit may reduce to 1/4 × 1/4 in. on thin film, a 16:1 reduction in area.

Thin-film circuits can be step and repeated onto a larger thin-film substrate. Typical substrate sizes are 3 × 3 in. allowing many individual circuits to be processed simultaneously. For example, there could be a 10 × 10 array of the 1/4 × 1/4 in. circuit mentioned above (allowing 1/4 in. of unused periphery for edge effects). Thus, although the processing cost is high, the individual circuit cost would be about 1/100 of the total cost; a significant reduction. Although the cost usually does not come down as low as the equivalent thick-film circuit, there may not be a significant difference, particularly for small production volumes.

The 3 × 3 in. thin-film array can be diamond or laser scribed and broken up into individual circuits. The scribing can be done before or after processing, and the scribed substrate is often termed a *snapstrate*.

7
LASER TRIMMING OF HYBRID CIRCUITS

7.1 THE NEED FOR TRIMMING

During the 1970s there was a very large increase in the manufacture of thick- and thin-film circuits. However, even with the best processing methods and equipment, it has been impossible to produce resistors with as processed tolerances better than 10 to 20%. The majority of circuits require tolerances much better than this. Analog-to-digital converters and digital-to-analog converters, for example, require tolerances down to 0.01%. Because of this, film resistors are normally processed on the low side and are brought into tolerance by removing part of the resistor material. The old standby in the hybrid industry was air abrasive trimming. In this process the resistor material was removed by "sandblasting" the resistor with a fine stream of particles. Although this process could be semiautomated, it often became the bottleneck in the production process. Precision circuits also required better tolerances than the best air-abrasive process could produce. The air-abrasion process also suffered from problems of downtime due to clogging of jets and pipes, wear on X-Y tables and nozzles, and uncleanliness of the process.

An obvious replacement for air abrasion was the use of the laser. In this process, part of the resistor is vaporized by bombardment with a high-powered laser beam. The process has been completely automated, the laser and X-Y table being controlled by a minicomputer. The laser trimming process has proven fast, reliable, precise, and low cost, and is now the industry standard for any reasonable throughput of hybrid circuits. The only advantage of an air-abrasive system is that it is low in cost compared with a laser system. For example, an expensive air-abrasive system may be $20,000 compared with $100,000 for a computer-controlled laser system. However, this capital investment becomes less important if high volumes are required.

7.2 THE LASER TRIMMING PROCESS

The two types of lasers used for trimming are YAG (neodymium-doped yttrium aluminum garnet) or CO_2 (carbon dioxide). Both lasers are repetitively Q-switched, to provide high-power pulses at a 5- to 10-kHz pulse rate. The YAG type has proved more popular since it pulses more rapidly and yields smaller spot sizes, making it more accurate for precision trimming. The YAG laser operates at a wavelength of 1.06 µm and can be focused to a spot size of about 20 µm (less than 0.001 in.). The peak power in the beam is very high; about 20 kW; but because of its low duty cycle, the average power is only about 10 to 20 W.

Various techniques are possible for providing the relative motion between the substrate and the laser beam during trimming. Some commercial systems employ motorized X-Y motion of the substrate table and probes. In other systems the laser head is moved relative to the stationary substrate or the beam scanned using X-Y deflective mirrors. The highest-precision systems move the laser head using a system of linear motors. With this technology, step sizes down to 2.5 µm and speeds up to greater than 100 cm/sec are possible. In all systems safety precautions are always used to prevent laser radiation damage to operators' eyes. Most systems are monitored using closed-circuit television cameras. When the laser beam is focused on the resistor material, the extremely high power densities heat, melt, and vaporize a small portion of the resistor material. The removal of the resistor material at the focal point causes the net resistance to increase. The kerf cut by the laser is approximately the same size as the laser spot diameter (usually about 0.001 in.). This trim technique can produce both large and small resistor changes. Figure 7.1 shows some typical trim configurations. In Figure 7.1A the laser has fired a single pulse and removed a single spot of material approximately the same diameter as the laser spot. The other three cuts in Figure 7.1 are made by pulsing the laser repetitively and moving the spot along the indicated cuts. This produces a continuous kerf where the material is removed.

The glass content of the thick-film resistor melts during the trimming process and tends to reseal the edge of the resistor cut on cooling. This tends to give some environmental protection to the resistor and improves the resistor stability compared with the air-abrasive techniques. Thick-film resistors are usually trimmed through the encapsulation glaze on the substrate.

Thin-film resistors also tend to oxidize or anneal after trimming, giving a stable process. The resistors can also be trimmed through environmental coatings such as polyimide. Redeposition of the resistive material after trimming has not been found to be a problem, although some commercial laser systems employ a fine jet of gas to blow the vaporized material away. However, this is usually employed to prevent

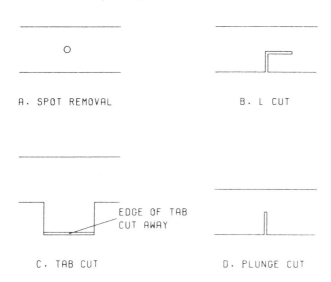

FIGURE 7.1. Typical laser trim configurations.

deposition of vaporized material on the expensive focusing lens rather than on the circuit. This absence of contamination due to trimming makes the laser trim technique particularly attractive for functional trimming with active components mounted on the substrate.

7.3 LASER TRIM SYSTEM

A typical laser trim system for passive trimming is shown in Figure 7.2. The probe board can usually hold sufficient probes to monitor up to 20 or 50 resistors. The laser is preprogrammed with the value of each resistor and its X-Y location. The substrate is loaded, aligned, and probed, and a start command is given to the computer. The laser beam is then directed to each resistor in turn and begins to trim. The scanner is switched to monitor the value of the resistor being trimmed and its value is measured in a very accurate digital voltmeter. As the resistor is trimmed, the measured value is continually fed back to the computer and compared with the required value. When this value is reached, the laser beam is switched off rapidly and the process is repeated on the next resistor. A major advantage of the process is that the laser trimming is very fast and the laser beam can be switched off within microseconds of the required value being reached. The laser trim time is usually only a few seconds for 10 or more resistors. The trim technique is a series of iterations of pulsing the laser, taking a measurement, making the comparison, and, if the value is still low,

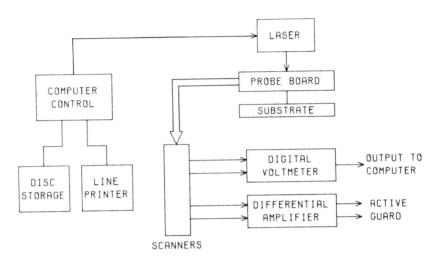

FIGURE 7.2. Laser trim system for passive trimming.

incrementing the laser position and repeating the process. This pulse-trim-measure routine takes time and can slow throughput on the laser. However, because of the predictability and accuracy of the process, and the use of a computer, some software routines can be used to speed up the process. Information is stored in the machine relating percentage trim to length of trim. After the first measurement the computer can now calculate how far to trim to get the resistor within a few percent of final value. Having done this, it will take a second measurement and again predict how far to cut. Using this method, only a few trim and measure commands may be needed to trim the resistor within tolerance. The throughput on the machine can be significantly increased.

Typical laser trimmers have measurement capability to ±0.005%, require very little software development for a particular circuit, and have trim times of 10 to 20 sec for even the most complex circuits. A limiting factor on the system may be the number of resistors that can be measured in one pass. The number of probes that can be mounted on the probe head without interfering with the trimming process is usually this limiting factor. Well-designed probe systems can accommodate about 30 to 40 probes, corresponding to simultaneous probing of 20 to 50 resistors. More complex resistor patterns may require further passes with different probe stations to trim the complete circuit.

Interfacing additional instruments to the system allows for functional trimming of the circuits. Accurate digital voltmeters allow trimming of dc gain, dc offsets, and other dc parameters, while further addition of waveform digitizers allows for trimming of ac parameters such as rise time and bandwidth.

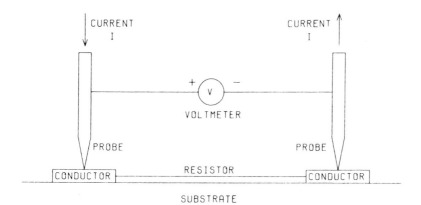

FIGURE 7.3. Two-point probe system.

7.4 RESISTANCE MEASUREMENT TECHNIQUES

Two-Point Probe System

A conventional two-point probe system for measuring a film resistor is shown in Figure 7.3. A known current is injected through probe pad 1, through the resistor and out of probe pad 2. The voltage across the two probes is measured.

Assuming that the voltage measurement is taken with a perfect voltmeter (infinite impedance), the equivalent circuit is shown in Figure 7.4. The resistance measured is

$$\frac{V}{I} = R_{film} + 2(R_{probe} + R_{contact} + R_{pad})$$

The probe resistance R_{probe}, the probe contact resistance $R_{contact}$, and the pad resistance R_{pad} may contribute from 0.1 Ω to a few ohms to the total resistance. Usually, the probe contact resistor is the

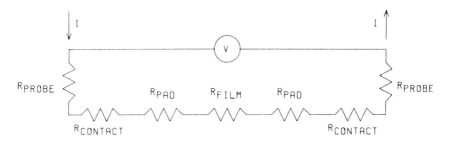

FIGURE 7.4. Equivalent circuit of two point probe system.

major problem. The value is not consistent from measurement to measurement, depending on the probe design, cleanliness of the contact pads, and pressure of the probe contact. This unwanted resistance can be a large percentage of low-valued resistors, resulting in trimming inaccuracies of the final resistor. For high-precision circuitry a four-point measurement is preferred over this two-point measurement.

Four-Point Probe System

A four-point probe system for accurate measurement of film resistors is shown in Figure 7.5. In this method a known current, I, is injected as before. However, two separate probes are used to monitor the voltage as close to the resistor edge as possible. The voltage monitor is very high impedance, such that the current in the voltmeter is many orders less than the current injected. The equivalent circuit is shown in Figure 7.6. With this system

$$V = I(2R_{pad2} + R_{film})$$

Thus using this technique the current probes and contact resistances are excluded from the measurement. The only error is the small series pad resistance, which may be less than one square of conductor, i.e., 5 to 10 mΩ. This technique is used to trim low-value resistors to very close tolerances.

A major problem is that four probes are required, double the number for a two-point measurement. Most probe cards can accommodate only 30 to 40 probes, and only 10 to 20 resistors can be measured at

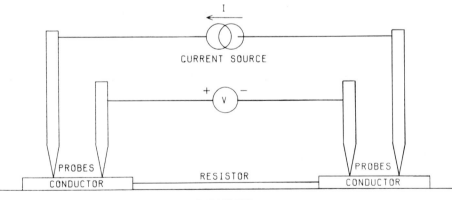

FIGURE 7.5. Four-point probe system.

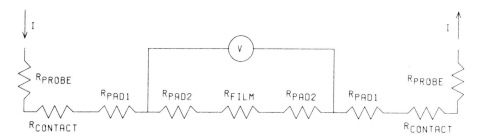

FIGURE 7.6. Equivalent circuit of four-point probe system.

one pass, compared with 20 to 40 resistors with a two-point measurement. The circuit should also be designed to accommodate the extra space needed for probing.

Most circuits are trimmed with a combination of two-point and four-point resistor measurements.

Closed-Loop Resistor Measurement

Resistor loops containing three or more resistors are often encountered in circuits to be trimmed. A typical circuit loop is shown in Figure 7.7.

If we probe Rx and trim it to value, the true value of Rx is masked by R1 and R2 in parallel. An active guard is usually provided in most laser trimming systems to mask out the effect of R1 and R2 while measuring Rx. The basic idea is shown in Figure 7.8.

The voltage at node 3 is picked off by a very high impedance voltage follower with gain very close to unity. The output voltage (also V1) is applied to node 2. Thus resistor R2 has equal voltages at its two end nodes and carries no current. Thus all the injected current flows through Rx, and resistors R1 and R2 are masked out. Usually, this masking technique does not require extra probes, since node 3 will already be probed to measure R1 and R2. The system does require the provision of the accurate voltage follower circuitry.

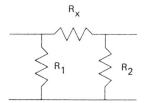

FIGURE 7.7. Circuit with resistors in a loop.

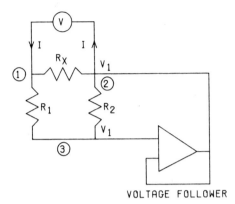

FIGURE 7.8. Active guard for trimming resistors in loops.

7.5 TRIM INFORMATION FOR TOP-HAT RESISTORS

The top-hat configuration shown in Figure 7.9 is one of the most popular trim configurations.

One problem when designing the layout for a top-hat resistor is computing the initial square count. Figure 7.10 shows the initial square count Ni as a function of the dimensions x, y, and w.

Example 1 Consider an example of a top hat with W = 40 mil, Y = 80 mil, X = 120 mil, and L = 120 mil. The square count for the top hat area is found from Figure 7.10. For Y/W = 2 and X/W = 3, Ni = 1.9 squares. The square count for the linear runs L on both sides of the top-hat region is 2L/W = 6 squares. Thus the total square count is 6 + 1.9 = 7.9 squares.

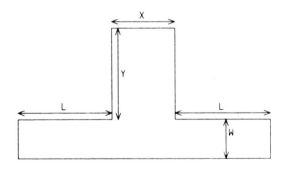

FIGURE 7.9. Top-hat geometry.

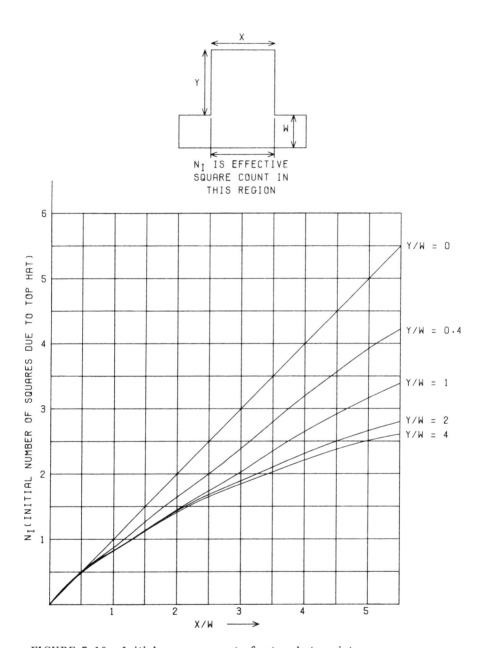

FIGURE 7.10. Initial square count of a top-hat resistor.

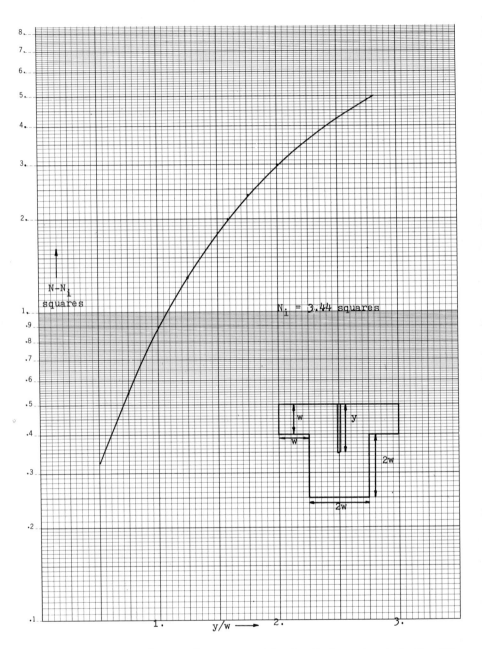

FIGURE 7.11. Resistor value versus trim depth for a top hat; X = 2W, Y = 2W, L = W.

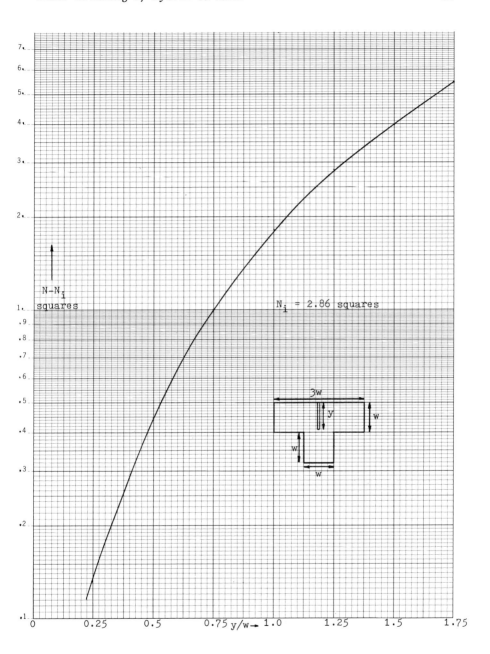

FIGURE 7.12. Resistor value versus trim depth for a top hat; X = W, Y = W, L = W.

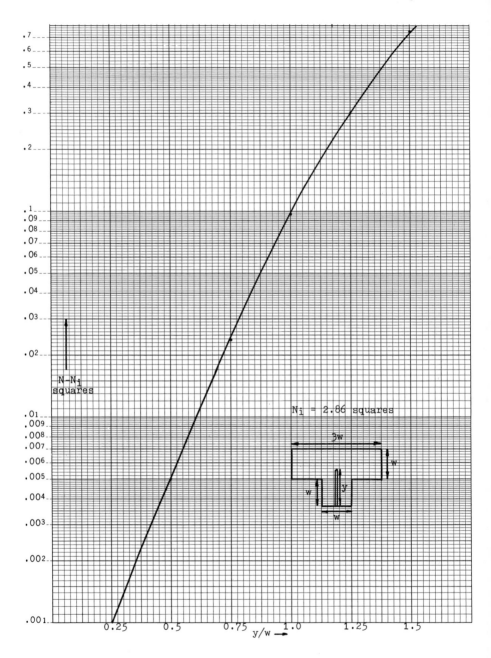

FIGURE 7.13. Resistor value versus trim depth for a reverse plunge cut.

Laser Trimming of Hybrid Circuits

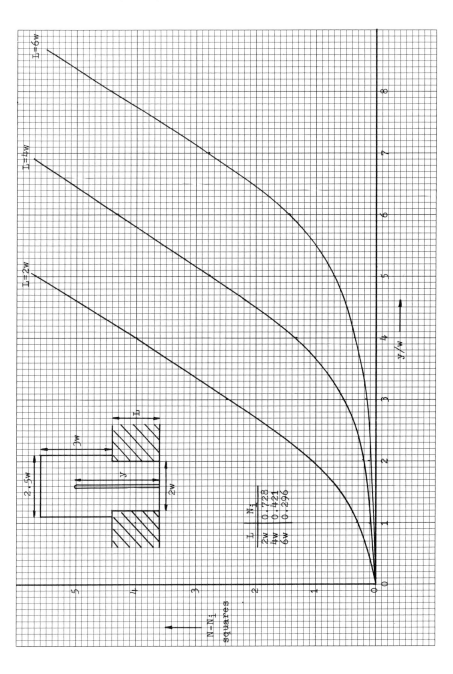

FIGURE 7.14. Resistor value versus trim depth for a subsquare top hat.

Example 2 Figure 7.11 shows the effect of trim depth for a resistor with Y = 2W, X = 2W, and L = W. From Figure 7.10 the initial value is 3.44 squares (1.44 squares for the top-hat area and 1 square each side of the top hat). The graph shows the difference in square count as a function of normalized trim depth Y/W. For example, if we trim to a depth of 0.8W, the additional square count is read from Figure 7.11 as 0.60 square. Thus the resistor value with this trim depth would be 3.44 + 0.60 = 4.04 squares.

Figure 7.12 shows a similar set of calculations for a resistor with dimensions Y = W, X = W, and L = W (half the top-hat width of Figure 7.11). The initial square count in this case is 2.86 squares. For a trim depth of 0.75W, the additional square count is 1 square, giving a total square count of 3.86 squares.

Figures 7.11 and 7.12 are very popular trim geometries for top hats. The results for top hats with the Y/W dimension greater than 2 can be calculated from these figures. Making Y/W greater than 2 does not significantly affect the initial trim count for the top-hat region. The result converges on 1.44 squares for X/W = 2 (Figure 7.11) and on 0.86 square for X/W = 1 (Figure 7.12). After a trim depth of 3W on Figure 7.11 the curve of N-Ni versus Y/W is essentially linear with a slope of 2. Similarly, after a trim depth of about 1.75W on Figure 7.12, the curve is also essentially linear with a slope of 4.

Figure 7.13 shows the results of an alternative method of trimming a top hat—the reverse plunge cut. For this type of cut the square count starts off extremely insensitive to plunge depth, which may be useful in some cases. Figure 7.14 shows the results for a subsquare top-hat resistor. The graph also shows the initial square count for various geometries.

8
ASSEMBLY TECHNIQUES

8.1 INTRODUCTION

Hybrid circuits range from simple resistor and resistor capacitor networks, to very complex hybrids with many different components and package configurations. Many different assembly processes are necessary to produce these complex hybrids. Some of these assembly operations will be similar to, if not identical to, techniques used in printed circuit board and integrated circuit assembly. However, some components and processes are unique to hybrid assembly. This chapter discusses the techniques and equipment necessary for the assembly of hybrids.

8.2 COMPONENTS USED IN HYBRID ASSEMBLY

Almost any component-type capable of being attached by soldering, epoxy, wire bonding, or equivalent process could be utilized in hybrids. Generally, leaded discrete components should have their leads formed in such a manner as to achieve planar mounting. This differs from conventional printed circuit board assembly, where the leads are inserted through the board. Unpackaged components such as semiconductor die, chip resistors, and capacitors are widely used. However, many potential component types are not economically or volumetrically compatible with the packing densities normally desired. Thus components such as transformers, relays, and potentiometers are not normally encountered in hybrid design.

Discrete Passive Devices

Resistors

Discrete resistors employed in hybrid circuits can be the usual axial leaded package or in a chip configuration. The chip resistor type is more compatible with hybrid circuitry as they offer good volumetric efficiency and are easier to attach using automated methods. The metallization on the chip resistor may be either thick or thin film.

Normally, resistors would be deposited in thin film or screened in thick film. However, there may be several reasons to employ chip resistors.

1. A design using thick film may have one resistor of such a value that a separate additional resistor pass is required. Use of a discrete resistor could be economically justified.
2. A design using thin film with low sheet resistivity may require a high-valued resistor. A discrete resistor may take up less space than a meander thin-film resistor.
3. Thick-film resistors have relatively high noise levels. A low-noise stage in the system may require the use of low-noise thin-film chip resistors.
4. Thick-film resistors have relatively high temperature coefficients of resistance. Temperature-sensitive circuitry may require lower-temperature-coefficient thin-film resistors.
5. Thick-film resistors do not have as good long-term stability as do thin-film resistors. Discrete thin-film resistors are used where stability may be a problem.

Capacitors

Discrete capacitors are used extensively in hybrid circuitry, both thick and thin film. Conventional disk, Mylar, and tantalum are used, but the most common type is the monolithic chip capacitor.

Although film capacitors are often used as an integral part of the circuit, they have a limited range, utilize a significant amount of substrate area, and require extra screening operations. The monolithic chip provides capacitances from a few picofarads to a few microfarads in a relatively small volume. Tantalum chip capacitors are used where high capacitance values are required.

Inductors

Discrete inductors can be attached to the hybrid in the same manner as discrete resistors and capacitors. However, larger inductor values above a few microhenries are not usually compatible with hybrid circuitry due to their large physical size.

Smaller values of inductance in the range 10 to 100 nH can be designed as planar conductor spirals. A typical film inductor is shown in Figure 8.1.

Assembly Techniques

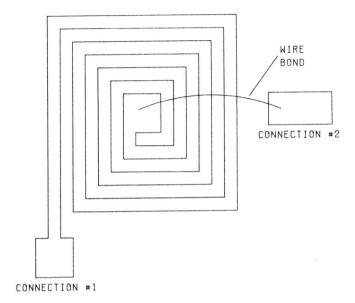

FIGURE 8.1. Film inductor.

Trimmers

Both trimmer capacitors and resistors can be solder reflowed to substrates. These can be used for "tweaking" circuit parameters as in conventional printed circuit boards. An alternative to attaching trimmers is to laser trim the resistors and capacitors with the circuit in a functional mode (called "functional trimming").

Active Devices

Transistors, diodes, integrated circuits, and almost every other type of active device have found some application in hybrid circuits. The only restriction on the type of active device that can be used is power dissipation.

Active devices suitable for hybrid circuitry are available in many different package styles.

Conventional Leaded Packages

These packages include standard transistor packages (TO5, TO92, TO39), dual-in-line ceramic or plastic IC packages, flatpack IC packages, and axial leaded diodes.

Leads are normally sheared and formed to be compatible with the normal planar mounting techniques.

Semiconductor Dice

Unpackaged chips (diodes, transistors, and ICs) can be attached to hybrid circuits using essentially the same processes as the semiconductor manufacturer uses to fabricate packaged devices. Semiconductor dice are used where maximum space economy is required. Also, unpackaged semiconductor dice exhibit better high-frequency parameters than the equivalent packaged device. Almost any commercially available semiconductor device is available in this unpackaged form.

Leadless Inverted Devices (LID)

The LID is a very small package for diodes, transistors, and ICs. The semiconductor is attached in the trough of the LID and wire-bonded to the legs (connections) of the device. The configuration is shown in Figure 8.2.

The semiconductor die is protected with an environmental coating and the LID is soldered to the hybrid. Although LIDs allow good space utilization, device availability is limited. Because of this, LIDs are not widely used in commercial hybrid circuits.

Flip Chips

The flip chip is a passivated semiconductor chip which has small solder bumps on its termination pads. The passivated surface gives environmental protection to the device. Figure 8.3 shows a typical device configuration.

The chip is placed with its active side next to the substrate and the solder bumps in contact with the metallization. Solder flux is used to keep the chip in place. The device is therefore "flipped" compared with normal die attachment—hence the term "flip chip." The substrate is heated and the die is soldered to the substrate.

Flip chips offer a high packaging density at a reasonable cost. A fairly wide range of devices is available and this type of device could gain in popularity.

Beam Lead Devices

This device is a passivated semiconductor chip with termination leads (usually gold) built up by plating techniques along the outer periphery of the chip. The devices are very expensive compared with the equivalent semiconductor die, and have not found wide application in hybrids. Major applications are in high-reliability military hybrids and microwave hybrids. A wide range of microwave transistors and diodes are available in beam lead packages.

Assembly Techniques

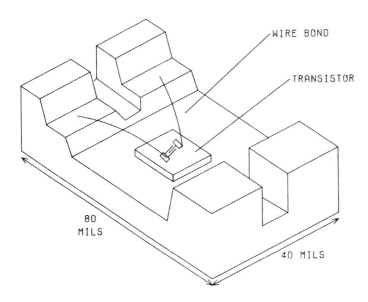

FIGURE 8.2. Leadless inverted device.

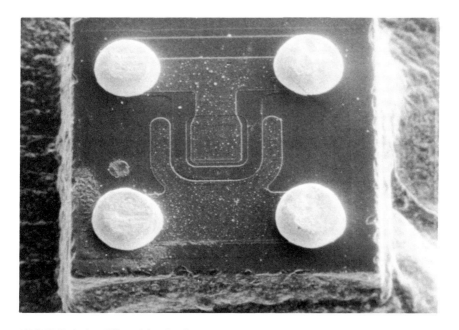

FIGURE 8.3. Flip-chip device.

FIGURE 8.4. SOT transistor and IC packages.

Micropackages

Several very small packages for semiconductor devices have been designed over the years. These include Satan, Minimold, Micro-e, and SOT (solder on transistor). Most of these devices have found limited application in commercial hybrids. The SOT device is gaining in popularity and is proving economical for commercial hybrids. Diodes, transistors, and ICs are available in the SOT package, and device availability is improving. Figure 8.4 shows SOT transistor and IC packages. The devices are reflow soldered to the hybrid metallization.

Assembly Techniques 101

Chip Carriers

Chip carriers are small multilayer ceramic packages with lead configurations up to 64 connections. They differ from dual-in-line packages in that the external connections are metal (usually gold) pads deposited on the ceramic. The semiconductor device is bonded within the package and a ceramic lid is usually soldered on to form an hermetic seal. Typical chip carrier configurations are shown in Figure 8.5.

The advantage of the chip carrier is that it is significantly smaller than the dual-in-line packages and allows for pretesting and burn-in of the device prior to attachment to the hybrid. The package is usually reflow soldered to the hybrid substrate. A disadvantage of the chip carrier is its relatively high cost compared with other approaches, although some of this cost may be recovered due to higher yields during final hybrid assembly. These higher yields are due to the pretest and burn-in capabilities.

External Connections

Hybrid circuits usually need some sort of lead attachment to interface with external circuitry.

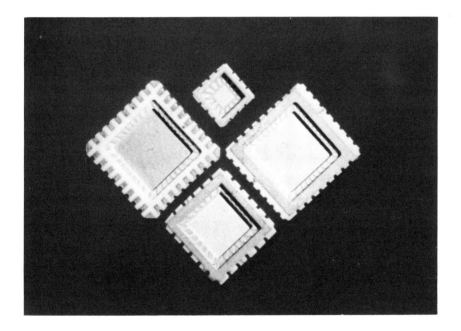

FIGURE 8.5. Chip carriers.

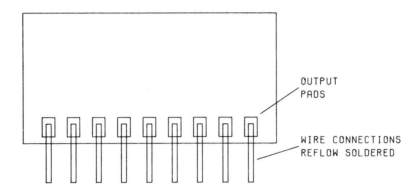

FIGURE 8.6. Wire connections.

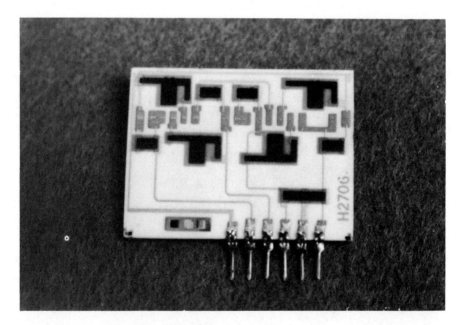

FIGURE 8.7. Terminal connections.

Assembly Techniques 103

Wires

Precut wires can be soldered to the substrate as shown in Figure 8.6. Some tooling is required to ensure accurate placement of the wires and to hold them in place during reflow soldering. The overall cost of this type of connection is very low.

Terminals

Several manufacturers offer a wide range of clip-on in-line and 90° terminals on various centerline-to-centerline spacings. Typical terminal costs are 1 to 5 cents per connection. The terminals can be attached in the appropriate lengths and solder reflowed to the substrate. The overall cost is very low, and this type of terminal is usually preferred over the wire terminations described above. Figure 8.7 shows some typical terminations.

Edge Connectors

Plug-in connectors are available which utilize finger contacts on the thick-film substrate. The connection is similar to the edge connectors encountered in printed circuit board design. This type of connection requires more space than do terminal connections. However, no soldering is required and hybrids can be easily replaced.

8.3 ASSEMBLY TECHNIQUES

Soldering

Soldering is probably the assembly operation most often encountered in hybrid circuits. Soldering is used for discrete device attachment, lead attachment, flip chips, and final package assembly and sealing operations. With appropriate materials and reasonable process control, soldering provides an economical and reliable method of assembly.

Solder Materials

Most solders in common use for hybrid circuitry are composed of alloys of tin (Sn), lead (Pb), and silver (Ag). Table 8.1 shows the properties of the most commonly used alloys.

Solder Fluxes

The solder fluxes used for soldering hybrids are similar to those used for printed circuit board assembly. Most metallization systems used have good soldering properties and only relatively mild fluxes need to be used.

TABLE 8.1. Common Solder Alloys

Sn (%)	Pb (%)	Ag (%)	Liquidus (°C)	Common metallization
60	40	—	238	PdAu
60	38	2	227	Ag, PdAg
10	90	—	302	PdAg, PdAu
90	10	—	213	PdAu
95	—	5	295	PdAg, PdAu
92.5	5	2.5	280	PdAg, PdAu

Soldering Processes

Thick-film substrates are usually immersed in molten solder to tin metallization pads prior to attachment of the components. Wires or terminals are usually attached at this time. This is particularly true if the clip-on type of terminal is used, as it is difficult to attach and maintain alignment after solder tinning due to the solder meniscus buildup on the pads. This pretinning process can be done by manual immersion in a solder pot, usually through an oil blanket, or in a more mechanized manner with a flow solder system.

Depending on the type of component to be attached, it is often desirable to control the solder buildup on a termination pad. This is particularly true when components such as flip chips and small chip capacitors and resistors are used. This is accomplished by pad design, the use of solder dams, and close control of the tinning procedures. Solder dams prevent the solder from running down a metallization conductor away from a termination pad. A good design of a solder pad using a solder dam is shown in Figure 8.8. With thick film, solder dams can be dielectric films, resistor material, or organic solder resists. Polyimide has been used for thin-film circuits. It is often desirable to cover the entire hybrid with solder resist except for the solder pads. This reduces failures due to solder shorts, reduces the quantity of solder used, and improves inspection procedures.

Solder Reflow Process. With a solder reflow process the components are placed in position usually in the presence of a flux, and the combination is heated to the reflow or melting point of the solder on the tinned substrate. Additional solder is not used, although the components may have also been pretinned. Generally, the substrate is preheated at a reduced temperature to avoid undue thermal shock. Various mechanisms are utilized to apply heat to accomplish the reflow attachment. Some of these methods are indicated in Table 8.2.

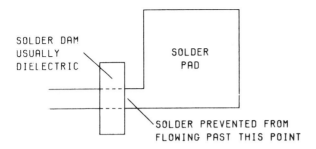

FIGURE 8.8. Solder dam.

Nitrogen or other inert or reducing atmospheres are used to prevent oxidation and promote more reliable solder joints.

Potential problems include metallization burn-off due to excessive time at high temperature, and cold solder joints if time and temperature are too low. The process therefore needs reasonable operator skill and good process control. The type of equipment utilized will depend on volume, with equipment costs of less than $50 for a hand soldering operation to $10,000 or more for automated systems.

Hand Soldering. This is an entirely operator-dependent operation and is usually for very low volume. Components are hand soldered with solder transferred to the joint with the iron, usually using wire core solder. The process requires a high skill level and is therefore usually high in cost. The process is more suited to a rework process wherein previously reflowed hybrids with failed solder joints are resoldered by hand.

Solder Paste. Solder paste or cream is finely ground solder mixed with fluxing agents and organic vehicles to make a screenable material. It is very compatible with the thick-film screening process. The material can be screened to the required thickness by selection of the

TABLE 8.2. Solder Reflow Processes

Soldering iron
Hot plate
Infrared heating
Hot gas
Reflow kilns (belt furnaces)
Vapor reflow

---

appropriate mesh size and by control of the screening parameters.
The paste is applied only to the areas to be soldered and in the quantities needed: the process therefore makes economical use of the materials. For low-volume circuits the paste can be applied using a dispenser or syringe.

Component leads are placed in the wet, tacky paste, which serves to hold the components in place prior and during the reflow operation. Some of the disadvantages of solder paste are:

1. Solder paste is tacky when screened, and handling of the screened substrate is a problem. Due to this handling problem, application is usually limited to one side of the substrate.
2. Components are positioned while the paste is tacky, and this can result in production flow problems.
3. The solder pastes tend to be more expensive than other soldering processes.

Cleaning After Soldering. After soldering, flux residues and other potential contaminants are removed by solvent cleaning. As thick-film substrate and films are generally impervious to attack by most common solvents, the selection of a cleaning solvent is dictated by the attached components, economics, and environmental restrictions. Trichloroethylene is widely used as a cleaning solvent. Vapor degreasers and ultrasonic degreasers are used to ensure reliable cleaning.

Chip-and-Wire Techniques

The term chip-and-wire refers to the attachment of semiconductor die to the substrate, and wire bonding of these devices to make substrate-chip interconnections.

Chip-and-wire requires specialized equipment and a high technical support level. It is therefore fairly high in cost compared with other assembly operations, such as soldering. However, the process gives maximum-density packaging, good reliability, and almost all available ICs can be obtained in chip form.

Eutectic Die Attach

Silicon and gold form a eutectic bond at a temperature just above 370°C. High-purity gold metallization is required, so the process is compatible with both thick and thin films. The bond is made by scrubbing the silicon die against the metallization at a temperature of 370°C. Equipment can range from a small hot plate with the die being hand scrubbed with tweezers, to more sophisticated bonders with vacuum die pickup, die substrate alignment mechanisms, and built-in ultrasonic scrub.

Assembly Techniques

The eutectic die attach method is identical to the method used in semiconductor packaging. However, with a hybrid circuit, several dice may need to be attached. This results in a major problem in that the first chip attached continues to eutectic flow as subsequent chips are attached. This can destroy the device and/or metallization if time is sufficient. Localized hot-spot bonders avoid this by utilizing hot gas jets, heated die collets, or focused infrared to heat only the die attach area. This localized heating is also required to replace defective dice during rework cycles.

Epoxy Die Attach

In many applications epoxy cements can be used to attach the die to the substrate. Both conductive and nonconductive epoxies are commercially available. The epoxy is placed using hand-held dispensers or using automatic epoxy dispensers which deliver exact amounts of epoxy to preprogrammed positions. The semiconductor dice are then placed in the epoxy, care being taken not to contaminate the active areas of the die. Some heat is required to cure the epoxy, most epoxy requiring 1/2 to 1 hr at 100 to 150°C. The advantage of epoxy attachment is that only moderate heat is required and devices can be removed and replaced relatively easily. The process is not suitable for thermocompression wire bonding, for which high temperatures are required, as the epoxy tends to lose its properties above about 200 to 250°C. The process is often used where room-temperature ultrasonic bonding is the wire bonding process.

Thermocompression Wire Bonding

Connections must be made from the metallization on the semiconductor die to attach to the metallization on the substrate. These connections are made using very fine wires, usually 0.7 to 2 mil in diameter. The process is called wire bonding.

Many metals exhibit the phenomenon of forming a metallurgical bond to like or different metals at a temperature considerably below the lower melting point of the two metals when pressure is applied normal to the point of contact. Thermocompression (TC) wire bonding is based on this phenomenon.

Usually, the metal on the semiconductor die is either gold or aluminum. The wire bonding materials suitable for these metal schemes are also gold and aluminum. However, the thin layer of oxide present on aluminum wire in air precludes its use for thermocompression bonding. Thus gold wire is the standard wire used in TC bonding process.

The gold wire is fed through a capillary, usually made of tungsten carbide, and a small hydrogen flame is used to cut the wire and form a ball on the end of the wire. At the start of the bonding sequence,

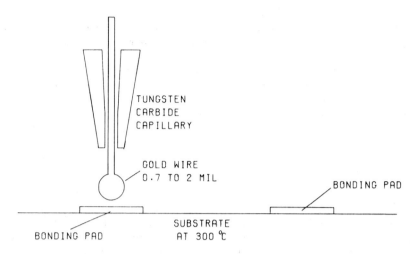

FIGURE 8.9. Initial setup for thermocompression wire bonding.

the setup is similar to Figure 8.9. In most cases the substrate is heated to 300°C to obtain the proper bonding temperature. If many wire bonds are to be made, this 300°C temperature for a long bonding period may degrade the semiconductors, and in this case the capillary is heated to about 350°C and the substrate lowered to about 200°C.

The ball and capillary are placed over the bonding area and the capillary and ball lowered onto the bonding pad. About 100 g of pressure is applied to the ball and bonding takes place. This "ball" bond is usually made on the semiconductor die. The capillary is then raised and moved to the termination metal on the substrate (usually thick- or thin-film gold). The wire has no ball in this case and the next bond is made by the edge of the capillary and is called a "stitch" bond. The capillary is then raised with the wire clamped. The wire is tagged away from the stitch bond and the hydrogen-flame-off jet is moved across to meet the wire, forming a new ball. The completed bond is shown in Figure 8.10.

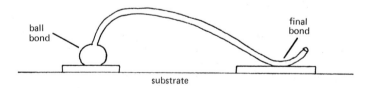

FIGURE 8.10. Final thermocompression wire bond.

Assembly Techniques

As an alternative to "gold ball" TC bonding, a small tungsten carbide wedge can be used. Both bonds are made using the wedge. This type of bond is preferred when bonding to very small, closely placed pads with centers in the range 1 to 2 mil. Microwave devices also have very small bonding pads and wedge bonding is often used in this application.

A major advantage of TC bonding is that bonding can be made in any direction after the ball has been bonded. Thus bonding can be accomplished over the entire substrate surface without rotating the substrate.

The major disadvantage of TC bonding for complex hybrids is the high temperature required for fairly long periods. This can result in drift in the parameters of both active and passive (film) components.

The equipment required for TC bonding is fairly complex and can cost $5000 to $25,000. The process requires careful contamination control procedures, a good preventative maintenance program, and skilled personnel.

Ultrasonic Wire Bonding

Ultrasonic wire bonding usually involves the use of aluminum wire, although gold can also be used. The wire is fed through and under a bonding wedge as shown in Figure 8.11.

During bonding the wedge presses the wire against the metal termination pad and ultrasonic energy (usually at about 20 to 60 kHz) is applied to the wedge. The wire is rubbed against the contact, causing local heating and a metallurgical weld. The thin oxide coating on aluminum wire is broken through and the oxide tends to help the friction heating process, giving a very reliable bond. Wire diameters

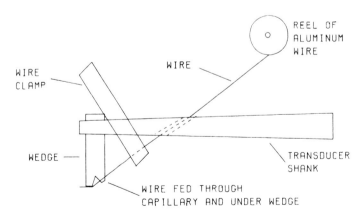

FIGURE 8.11. Ultrasonic wire bonding.

used are similar to TC bonding, typically 0.7 to 2 mil. After the first
bond has been made, the substrate is moved relative to the wedge,
pulling the wire through the hole in the wedge. The substrate movement can be in one direction only, in the direction of the wire feed.
The substrate is positioned until the wire is over the second termination
pad and the process is repeated. On completion of the second bond
the wire is clamped and tagged away from the bond, leaving a short
tail.

Ultrasonic bonding has several advantages: no heated substrate
stage or heated capillaries are required, and aluminum is more economical than gold.

A major disadvantage is that the substrate has to be rotated after
each bond to set up the direction for the next bond. For hybrids
with many devices the disorientation caused by many rotations can
lead to frequent bonding errors. Equipment costs are in the range
$4000 to $20,000. The process requires careful control of the ultrasonic energy coupled to the bonding tool, the pressure applied to the
bond, a good preventative maintenance program, and skilled, careful
operators.

Beam Lead Attachments

Beam lead assembly offers an alternative to the chip and wire operations
described above. The beam lead alternative requires additional processing at the wafer level to create the beam leads, and a beam lead attachment technique. Figure 8.12 shows a typical beam lead structure.
The beam leads are small gold bars cantilevered from the semiconductor
die. The beam is typically 0.0005 in. thick, 0.002 to 0.004 in. wide,
and 0.010 to 0.015 in. long. The beam is a gold lead with thin layers
of platinum, titanium, and sometimes chromium deposited on the semiconductor side. The beam is insulated from the body of the device
by a layer of oxide or nitride. This insulation layer extends over the
beam for 1 or 2 mil.

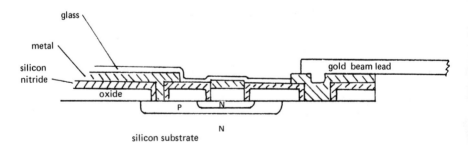

FIGURE 8.12. Beam lead structure.

Assembly Techniques

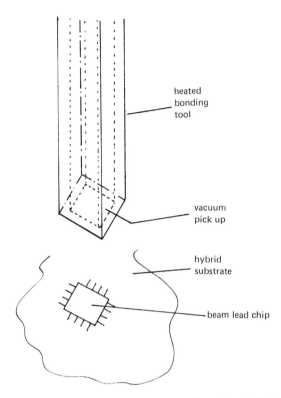

FIGURE 8.13. Beam lead bonding (wobble bonding).

The beam leads are thermocompression bonded to gold termination pads on the heated substrate. Using a tool similar to that shown in Figure 8.13, all the beam leads can be bonded at one time.

In practice, thickness variations and lack of exact parallelism between tool and substrate makes direct pressure unworkable. However, if after contacting the beam lead device, the tool is tilted to one side and the direction of tilt swept through 360°, a high bond yield is achieved. The process is called "wobble bonding." Since only a portion of the tool is in contact at any time, the effects of thickness variation and parallelism are minimized.

Although bonding economy is achieved due to simultaneous bonding of all leads, the cost savings are usually lost because of the high initial die cost. The major advantage of beam lead is the high strength of the bond, resulting in a very reliable hybrid.

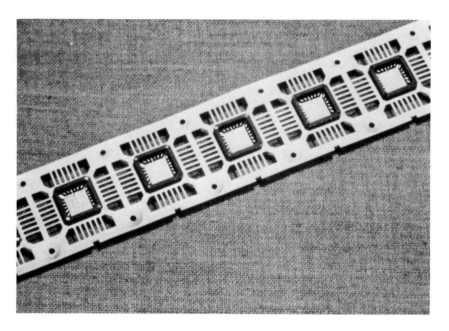

FIGURE 8.14. Tape carrier.

Tape Carrier Bonding

The tape carrier bonding system is an established system for automated interconnection of individually packaged semiconductor devices. Individual dice are bonded to lead frames etched along a strip of polyimide tape. A typical carrier is shown in Figure 8.14.

Three processes are required for a tape carrier system. The first is the fabrication of the tape carrier and lead frame. The tape is most often a polyimide film with copper laminated to the surface. The lead frame pattern is created by photolithographic processing and etching. Figure 8.15 shows a typical tape cross section. The second process required is the gang bonding of the semiconductor die to the lead frame. This process is called *inner lead bonding* and utilizes either thermocompression or solder joining techniques. This operation is a critical part of the process and requires compatible metals on the lead frame and semiconductor die. The third process is the gang bonding of the devices to metallized ceramic substrates. This process is called *outer lead bonding* and consists of severing the chip and part of the lead frame from the carrier, forming the severed leads and thermocompression bonding of these leads to the substrate. The final assembled chip is shown in Figure 8.16.

Assembly Techniques 113

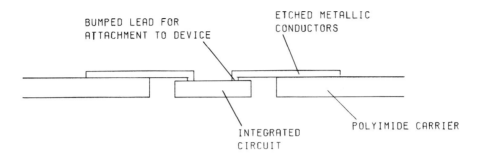

FIGURE 8.15. Tape carrier cross section.

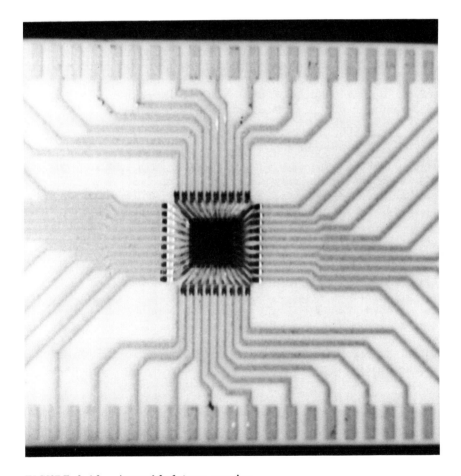

FIGURE 8.16. Assembled tape carrier.

The tape system offers three major advantages over other wire bonding methods. The tape system provides a means of fully electrically testing individual semiconductor chips prior to committing them to the substrate. This capability provides a significant cost savings by greatly improving first-pass yields and reducing rework. This cost savings usually outweigh the extra cost involved in the more complex processing required. A second advantage of the tape system is the potential to burn in semiconductor chips on tape, prior to assembly. This improves reliability of the assembled hybrid by reducing infant mortalities. The third major advantage is that the pull strength of a tape bond is approximately 50 g, compared with about 5 g for a TC wire bond. This increases the reliability of a tape-bonded hybrid.

One disadvantage of the tape method is the poor availability of pretaped semiconductors from vendors. To take full advantage of the method, hybrid manufacturers need to extend their capability and tape their own devices. This involves purchase of inner and outer lead bonders and the development of plating processes to plate appropriate metallization on the bonding pads of commercial wafers. A significant capital expenditure is therefore required to set up a tape-bonding facility. However, for large volumes of complex multichip hybrids, tape bonding should be seriously considered.

Automated Wire Bonding

Automated thermocompression bonding eliminates individual operator adjustments by automatically bonding all the wires between the chip and the substrate, according to a preprogrammed sequence. A typical automated wire bonding system consists of a computer-controlled bonding head and high-speed X-Y table, a closed-circuit TV targeting system, a computer programming capability, and a feed mechanism for loading, indexing, and unloading the hybrid substrates. Usually, the chips are aligned manually by the operator and the bonding is done under computer control. Some attempts are being made to include pattern recognition systems to align the chips automatically. Most automated wire bonders can operate at 5000 to 10,000 wires per hour compared with about 500 to 1000 per hour for manual bonders. Significant savings can be obtained with such a system. Elimination of operator control also results in fewer bonding errors and more consistency in bonding quality. Automated bonding systems are fairly expensive to set up, typically in the $50,000 to $100,000 range. However, for high-volume hybrid assembly, significant overall savings can be made.

Conductive Epoxy

Conductive epoxies filled with silver (occasionally, gold) are frequently used to achieve component attachment without the use of solder. The major advantage of conductive epoxy is that it is a room-temperature-application process. Single-component systems require no mixing and have curing temperatures of 150 to 250°C. Two-part epoxy systems require mixing and have a typical curing temperature of 100°C. Almost any component that can be attached by soldering can be attached with conductive epoxies.

The epoxy can be dispensed manually with a hypodermic syringe or with an air-pressure-controlled syringe dispenser. Sufficient epoxy must be used to ensure good electrical contact and to provide adequate mechanical strength. The conductive epoxies are considerably more expensive than solder and generally require more labor to apply. The process is usually used only for low-volume assembly or where temperatures above 150°C cannot be tolerated.

8.4 PACKAGING

After completing assembly operations such as soldering, chip attach, and wire bonding, it is usually necessary to provide some sort of protective packaging. This can range from expensive hermetic packages, molded packages, and conforming dip coatings to occasionally nothing at all.

Hermetic Packages

Hermetic packages are used in high-reliability-type applications where maximum environmental protection is required. Commercially available hermetic packages are available in metal, glass, or ceramic. Usually, the hybrid substrate is mounted in the package cavity after all bonding operations have been completed. The substrate is often soldered or epoxy attached and interconnected to the package leads by wire bonding techniques. The package cover is attached by soldering, welding, or glass fusion, usually in an inert atmosphere. The quality of the hermetic seal is specified by the leak rate (e.g., 10^{-8} cm^3/sec). On typical hybrid packages exceeding 1 × 1 in. leak rates better than 10^{-8} cm^3/sec are difficult to achieve.

Hermetic packages are very expensive, often exceeding the cost of the hybrid substrate and components. Hermetic sealing equipment is also expensive, requiring inert atmosphere kilns, welding and brazing equipment, and helium-leak detectors to qualify the degree of hermeticity.

Ceramic Covers

Often, the only components requiring protection on the substrate are the semiconductor dies and their associated wire bonds. The most common method of protecting these is to enclose them under a low-cost ceramic cover as shown in Figure 8.17.

FIGURE 8.17. Ceramic covers.

The ceramic cover is usually attached using a preform of nonconductive epoxy, and provides the semiconductors with both mechanical and environmental protection. The process requires almost no capital equipment, is low in cost, and provides a reliable method of protection.

Conformal Coatings

For high volumes conformal coating is probably the most common protective method. Materials utilized include liquid coatings such as epoxies, urethanes, phenolics, and diallic phthallates, which are generally applied by dipping, followed by a cure cycle. Several coats are often applied. If uncased semiconductors are being coated, a silicone or urethane precoating is often applied. These materials provide the semiconductor with a contamination barrier and also absorb any stresses on the wire bonds. Alternatively, the semiconductors can be protected with a ceramic cover prior to coating.

Material buildup on irregularly shaped hybrids is difficult to control, so overall package dimensions cannot be accurately specified on conformally coated packages. Phenolic coatings can be made impervious to moisture by vacuum impregnation with a high-temperature wax.

Epoxy and fluidized-bed materials are also often used. The component is heated and immersed in a cold fluidized bed of the epoxy material. The fluidized bed is a container of powdered material with air flowing through it. The powder has the appearance and properties of a liquid. During dipping, the powder melts and adheres to the part. On removal from the fluidized bed, the part has a very uniform coating of material on all surfaces. The material is then cured. Conformal coating equipment is relatively simple and usually ranges from about $2000 to $30,000.

Molding

Hybrids can also be molded in package sizes similar to integrated circuits (i.e., dual-in-line and single-in-line packages). Such packages have good size and uniformity and are compatible with automatic component insertion into PC boards.

The materials used are generally epoxies, silicones, and diallic phthallates. Glass or mica fillers are often used to help compensate for thermal coefficient differences. This is because alumina and epoxy have large differences in thermal expansion coefficients: $6.8 \times 10^{-6}/°C$ for alumina and $50 \times 10^{-6}/°C$ for epoxy. Transfer molding of the thermoset materials is usually used and parts are fabricated in strips to enable several parts to be molded simultaneously.

Molding equipment is fairly expensive, usually in the range $30,000 to $60,000. Also, individual molds cost in the range $4000 to $40,000, depending on complexity. This type of package is therefore used for high-volume standardized configurations.

Specialized Packages and Coatings

On low-volume applications a simple potting or casting procedure may be used with liquid epoxies or silicones. A low-cost metal mold may be used or a low-cost case may be merely filled with the encapsulant. Investment is very low but the labor cost is high, so the approach is used only for low volumes or prototypes.

A recent development is parylene vapor deposition which deposits a very thin but uniform coating evenly on all surfaces of the hybrid. A nearly hermetic coating can be obtained without any package. The equipment and process is available on a license basis only, although several contract coating sources are also available.

9
PRODUCTION CONSIDERATIONS

9.1 DOCUMENTATION

Documentation is a very important part of any manufacturing process. With a hybrid manufacturing capability, complete documentation should be kept on material, equipment, and processes.

Material Specifications

The quality and consistency of the materials used is one of the keys to a successful hybrid manufacturing operation. Critical properties of the materials should be specified, in as much detail as possible. Detailed procedures should be available for inspecting and testing incoming material. This incoming inspection is very important, since material problems should be isolated before issuing the materials to the production floor.

Process Specifications

Hybrid production involves many different processes, covering a wide range of technologies. Detailed process specifications are required to ensure that close control can be maintained on all the process variables. Detailed process documentation is also a very important aid in the training of personnel.

Documentation Package

Each hybrid design should have a complete documentation package that clearly defines the specific physical and electrical requirements of the device.

Typical Hybrid Specifications

The following is a list of the typical documentation necessary for a hybrid circuit.

   Substrate specification
   Component specification
   Bill of materials
   Engineering schematics (electrical schematic; assembly drawings)
   Mechanical schematics (package and final assembly)
   Product flowchart
   Process flowchart
   Reliability specification
   Laser trim specification
   Test specification

9.2 INSPECTION AND RELIABILITY

In-process inspection is necessary both to ensure that the final product meets its specifications and to act as a control mechanism on in-process yield losses. Although manufacturing has responsibility for processes, materials, and quality, a strong, independent inspection program serves as an additional check that all specifications are maintained. Incoming inspection of materials and components prior to release to the production lines should be part of this program. Materials should be checked against specifications and components should be checked both mechanically and electrically.

Typical inspection gates should occur at the following points in the process:

1. After screening or deposition, visual
2. After resistor trimming, visual and electrical
3. After assembly, visual
4. Final package, visual and electrical

Typical inspection criteria are given in Tables 9.1 and 9.2.

TABLE 9.1. Electrical Inspection Criteria

Absolute resistance value	Per print
Matching resistance value	If required
Absolute temperature coefficient of resistivity	If required
Matching temperature coefficient of resistivity	If required

TABLE 9.2. Visual Inspection Criteria

Pattern registration
Line widths
Line spaces
Scratches
Inclusions
Resistor/conductor overlaps
Laser trim locations
Depths of laser trims
Cleanliness of laser kerfs
Crazing of resistors (microcracks)
Substrate imperfections

9.3 TECHNICAL SUPPORT

The wide variety of processes and technologies required for the production of hybrids necessitates a fairly high level of technical support for manufacturing. Expertise will be required in the following areas:

1. *Materials*: Thick- and thin-film materials, photoresists, solders, encapsulants, epoxies, solvents, packages
2. *Processes*: Screening, film deposition, firing, laser trimming, soldering, chip and wire bonding, leak testing
3. *Tooling*: Screening, deposition, vacuum techniques, laser trimming, test jigs, final assembly
4. *Testing*: Component tests, diagnostic testing, laser trimming (passive and active), final test

In general, thin-film processes require more technical support than do thick-films processes. Areas such as testing and laser trimming may require a very high level of support to utilize the facilities fully. Laser trimming will require significant systems and software expertise to run and maintain the complex equipment required.

9.4 LABOR ASPECTS

Thick Film

Screening

Typical thick-film screening rates are shown in Table 9.3. A thick-film hybrid with three resistor layers, two conductor layers, and an encapsulant could be produced at an overall rate of about 50 per hour with manual equipment and about 250 per hour with automatic equipment. These figures do not include setup times for the screening

TABLE 9.3. Screening Rates

Equipment type	Substrates per hour
Manual laboratory equipment	200–300
Typical production equipment	300–600
Fully automatic equipment	500–1500

equipment. However, at high production rates the setup time is usually a small percentage of the overall time. These rates could be increased with multiple images on a master substrate.

Firing

Kiln loading and unloading contributes a negligible amount of labor cost.

Laser Trimming

Laser trimming can run up to about 15,000 resistors per hour, depending on design, tolerance, and loading/unloading times. The average substrate with about 20 resistors may require about 1 to 2 hr of initial software development for the passive trim program. Active trimming programs may require weeks of software development to develop an efficient trimming procedure. Efficient use of predictive trim algorithms is a necessity to obtain optimal return on the high capital investment in laser trim equipment.

TABLE 9.4. Assembly Rates

Process	Rate/hour
Manual wire bond	400
Automatic wire bond	3000
Die attach	150
Component attach	100
Lead attach	700
Inspection	250
Package seal	150
Leak check	500
Final test	100

TABLE 9.5. Assembly Yields

Process	Yield (%)	
Wire bond	98	
Die attach	99	
Component attach	99	
Lead attach	99.5	
Package seal	99	
Final test	75	Complex hybrid
	95	Simple hybrid

Assembly Processes

Table 9.4 shows typical assembly rates for hybrid production. These assembly rates will vary greatly with the degree of automation in the production process.

9.5 YIELDS

There are yield losses at every step in the process. These losses can make major contributions to the final part cost, and careful attention should be made to where and why these losses are occurring. Table 9.5 shows typical yields for some of the commonly used processes.

Generally, yields after trimming and test are from 70 to 90%, depending on the complexity of the component. Average daily yield is also a useful tool in monitoring the process control.

9.6 IN-HOUSE OR EXTERNAL MANUFACTURING

Most companies start out by purchasing hybrids from vendors or custom houses. However, when significant numbers of hybrids are being purchased, the question arises as to whether to establish a hybrid in-house capability. There are several advantages to in-house manufacture:

1. Better understanding and utilization of hybrid technologies.
2. Better specifications on the components designed due to understanding the limitations of the technology.
3. Component changes can be implemented quickly due to the short communications channel between design and production.

However, before setting up a facility, the following points should be analyzed:

Production Considerations

1. Does the potential volume justify an in-house facility?
2. Should the facility be thick film, thin film, or both?
3. Would hybrid technology be incompatible with other in-house facilities?
4. Are there any staffing problems?
5. Would the hybrid facility be compatible with existing components being purchased from hybrid vendors?
6. Should the facility start up as a prototype facility and lead in to full production?
7. Can a significant facility be established so that all the advantages of hybrid technology can be obtained?

10
THERMAL CONSIDERATIONS

10.1 INTRODUCTION

Advances in the complexity of hybrid circuits have yielded devices with many components, some of which may dissipate appreciable power. This has resulted in higher operating temperatures and an increased probability of component failure. Temperature limits the safe operating region of all electronic components. Above 125°C junction temperatures, transistors may suffer from thermal-runaway problems. Above temperatures of about 150°C, film resistors exhibit appreciable drift and may even fail catastrophically. For high reliability circuits, methods must be provided for the efficient removal of the heat created by these heat-dissipating components.

Efficient thermal design depends on understanding and analyzing the mechanisms of heat flow.

10.2 COOLING PHENOMENA

The basic problem of thermal design is to provide a path for the removal of thermal energy from a high-temperature heat source to a low-temperature heat sink. There are three basic cooling phenomena:

Conduction: This is the transfer of thermal energy from the high-temperature region of a solid to a lower-temperature region. The energy flow is through the solid.

Convection: This is the transfer of thermal energy from the high-temperature surface of a solid body to a lower-temperature surrounding liquid or gas. The usual method of convection cooling is by means of the surrounding air, although in some cases water cooling can be used.

Radiation: This is the transfer of thermal energy from the surface of a solid body to the ambient surroundings by electromagnetic waves.

Thermal Considerations 125

Generally, efficient thermal designs will attempt to make use of all three methods of cooling. Metal heat sinks (usually copper) will be used for conduction. Heat sinks may have finned extrusions to aid convection, and have dark-colored surfaces to emit radiant energy.

10.3 COOLING OF HYBRID MICROCIRCUITS

The amount of convection and radiation cooling of a hybrid is usually negligible, and most of the thermal energy is removed by conduction. The design must therefore provide a means of removing steady-state heat by conduction from components on the substrate to the package, case, or leads. However, because of the fewer thermal interfaces and the large thermal junction cross sections, higher power densities can be achieved in hybrids than in discrete components.

The heat flow is relatively simple, flowing from the heat source, resistor or transistor, through the substrate to a low-temperature heat sink such as copper strips on an etched circuit board or a finned extruded heat sink attached to the nonactive side of the substrate. The problem of steady-state analysis is fairly simple for most circuits. However, the problem of transient heat flow is much more complex and usually requires a computer simulation for adequate solution.

There are many "rules of thumb" around for the thermal design of hybrid circuits. These are mentioned here but should be used with caution; they may be very pessimistic in some cases and very optimistic in others. For all but the simplest circuits it is usually advisable to do a complete thermal study.

Rules of Thumb

1. A good rule of thumb for resistors is to design for less than 50 W/in.^{2}.
2. Usually, 125°C is a good figure for the maximum allowable junction temperature for transistors and integrated circuits. A good rule of thumb is that the IC failure rate is doubled for every 10°C increase in temperature.
3. To estimate temperature drop of a single-emitter transistor junction to the die pad, a useful approximation is

$$\theta_{j\text{-}p} = \frac{1}{4K\sqrt{LW}}$$

where

$\theta_{j\text{-}p}$ = thermal resistance from the junction to the pad
K = semiconductor thermal conductivity (K for silicon at 100°C is 1.1 W/cm-°C)

L = length of source of heat, cm
W = width of source of heat, cm

Usually, L and W can be approximated as the transistor emitter dimensions.

Although this approximation is partially successful in matching experimental data, it should be used with caution. This is particularly true with ICs, where there may be many individual sources of heat.

Example Estimate the temperature rise of a single-emitter transistor 16 µm long and 3 µm wide, dissipating 30 mW.

$$\Theta_{j-p} \equiv 330°C/W$$

Thus

emitter temperature - pad temperature = 330°C/W × 0.03 W

$$= 10°C$$

That is, the emitter will be 10°C above the substrate pad.

10.4 ANALYTICAL TECHNIQUES

The basic equation for the conduction of thermal energy is

$$Q = \frac{KA}{L} \Delta T = \frac{KA}{L} (T_1 - T_2) \tag{10.1}$$

where

Q = heat flow/unit time, W
K = thermal conductivity, W/cm-°C
A = area of the thermal path, cm^2
L = length of the thermal path, cm
T_1 = temperature of the heat source, °C
T_2 = temperature of the heat sink, °C
$\Delta T = T_1 - T_2$

The thermal conductivity is a measure of how well a material conducts heat. For example, a beryllia substrate, an excellent thermal conductor, has a much higher thermal conductivity than does an alumina substrate.

The conductivity equation can be written in the form

$$Q = \frac{\Delta T}{L/KA} = \frac{\Delta T}{\Theta} \tag{10.2}$$

where Θ is called the thermal resistance of the conductivity path and has the units °C/W. Rewriting the equation in the form of Equation (10.2) creates a thermal equivalent to the electrical equation known as Ohm's law.

$$I = \frac{V}{R} \quad \text{Ohm's law of electrical conduction}$$

$$Q = \frac{\Delta T}{\Theta} \quad \text{thermal conduction}$$

In these equations:

The term Q is analogous to current I.
The term ΔT is analogous to voltage V.
The term Θ is analogous to resistance R.

Using this electrical analogy one can analyze heat flow through a conductive path in a similar manner to the analysis of current flow through a resistive network.

Figure 10.1 shows the thermal and electrical analogies.

With a hybrid circuit, the problem often encountered is shown in Figure 10.2. Heat flows from a transistor heat source at temperature T_1 through the substrate to a heat sink at temperature T_2. The thermal path is through the silicon device, the eutectic bond, substrate, and the bond from substrate to the heat sink.

The thermal resistances of each component are in series and can be added algebraically in a similar manner to series resistors in the electrical analog. Thus

$$\Theta_{total} = \Theta_1 + \Theta_2 + \Theta_3 + \Theta_4$$

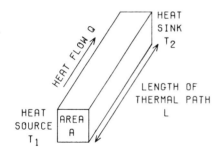
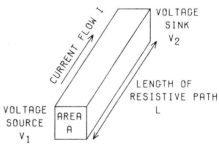

$T_1 - T_2 = Q \Theta$ $V_1 - V_2 = IR$

$\Theta = \frac{L}{KA}$ $R = \frac{L}{\sigma A}$

K = THERMAL CONDUCTIVITY σ = ELECTRICAL CONDUCTIVITY

A) THERMAL FLOW   B) ELECTRICAL FLOW

FIGURE 10.1. Thermal and electrical analogies.

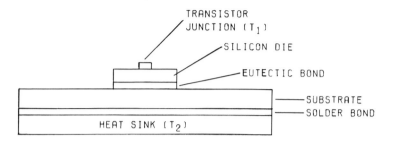

FIGURE 10.2. Hybrid thermal analysis.

where

Θ_1 = thermal resistance of the silicon device
Θ_2 = thermal resistance of the eutectic bond
Θ_3 = thermal resistance of the substrate
Θ_4 = thermal resistance of the bond from substrate to heat sink

and the solution for the transistor temperature above the heat-sink temperature is

$$Q = \frac{\Delta T}{\Theta_{total}}$$

where Q is the power dissipated in the transistor source.

Another often encountered hybrid structure is when the device is connected by several edge connections to an etched circuit board. This is shown in Figure 10.3.

In this case the thermal energy flows through the series path transistor-silicon-eutectic bond-substrate. There will now be several parallel heat flow paths to the etched circuit board: namely, the leads on the hybrid. The thermal resistances of each parallel lead can be added in the same manner as parallel electrical resistances:

$$\frac{1}{\Theta_{total}} = \frac{1}{\Theta_1} + \frac{1}{\Theta_2} + \frac{1}{\Theta_3} + \cdots$$

Thus for N identical leads, each with thermal resistance Θ_L, the overall thermal resistance due to the leads will be

$$\Theta_{total} = \frac{\Theta_L}{N}$$

and the overall system thermal resistance will be

$$\Theta_{system} = \Theta_{die} + \Theta_{eutectic\ bond} + \Theta_{substrate} + \frac{\Theta_L}{N}$$

Thermal Considerations 129

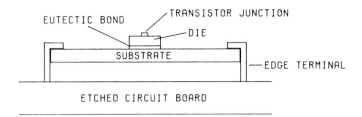

FIGURE 10.3. Hybrid with edge connectors attached to an etched circuit.

In the practical analysis of a thermal network, certain elements can be treated as isotherms; i.e., there is no temperature drop across the element, which gives it a thermal resistance of zero. Examples of hybrid elements that can be treated in this manner are film resistors and conductors. This is because they have thin cross sections and have intimate contact with the substrate, as shown in Figure 10.4.

Most intimate bonding techniques can also be considered as isothermal elements, e.g., eutectic bonds and thin solder bonds.

In practice, difficulties are encountered in using the thermal conductivity equations. This is because most heat sources have a small surface area (e.g., transistors) and are connected to larger surface areas (e.g., substrates). The thermal path tends to spread out from the source throughout the thermally conductive substrate. However, some reasonable approximations can be made:

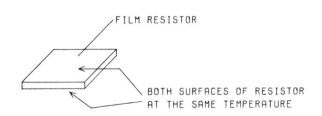

FIGURE 10.4. Film resistors and isotherms.

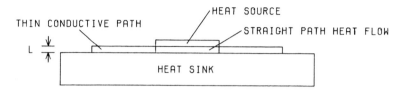

FIGURE 10.5. Heat flow through thin conductive path.

1. For thin sections compared to the heat source dimensions, one can assume that the heat travels in a straight path as shown in Figure 10.5. The thermal resistance of this path will be

$$\Theta = \frac{L}{KA}$$

where

    L = length of path
    A = surface area
    K = thermal conductivity

2. For thermally conductive paths that are thick compared to the dimensions of the heat source, a good approximation is to assume that the heat flows out evenly from the source at a 45° angle as shown in Figure 10.6. This increases the area of the thermal path, resulting in a much lower thermal resistance than that given by the straight-path calculation. The thermal resistance in this case will be

$$\Theta = \frac{1}{K} \int_0^L \frac{d\ell}{A}$$

where

    $d\ell$ = incremental length of path
    A = surface area at any cross section of heat flow

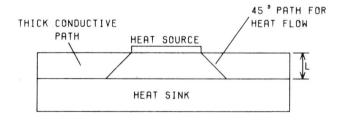

FIGURE 10.6. Heat flow through thick conductive path.

Thermal Considerations

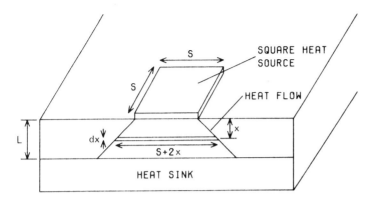

FIGURE 10.7. Square isothermal heat source.

Note: A is a function of distance from the heat source.

Since condition 2 is encountered often in hybrid design, some examples will illustrate its application.

Square-Shaped Heat Source

In this case a square isothermal heat source with sides of length S is mounted on a substrate with thickness L, as shown in Figure 10.7. The substrate is mounted on a heat sink. The thermal resistance between the heat source and heat sink is given by

$$\Theta = \frac{1}{K} \int_0^L \frac{dx}{A} = \frac{1}{K} \int_0^L \frac{dx}{(S+2x)^2}$$

$$= \frac{1}{K} \left[\frac{-1}{2(S+2x)} \right]_0^L$$

$$= \frac{1}{2K} \left(\frac{-1}{S+2L} + \frac{1}{S} \right)$$

That is,

$$\Theta = \frac{L}{KS(S+2L)}$$

Rectangular-Shaped Heat Source

In this case a rectangular isothermal heat source with sides S_1 and S_2 is mounted on a substrate with thickness L as shown in Figure 10.8. The substrate is mounted on a heat sink. The thermal resistance between the heat source and heat sink is given by

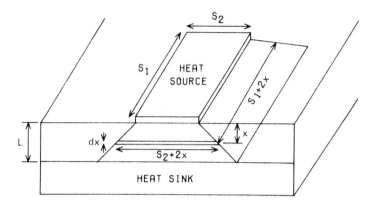

FIGURE 10.8. Rectangular isothermal heat source.

$$\Theta = \frac{1}{K} \int_0^L \frac{dx}{A} = \frac{1}{K} \int_0^L \frac{dx}{(S_1 + 2x)(S_2 + 2x)}$$

$$= \frac{1}{2K(S_1 - S_2)} \int_0^L \left(\frac{1}{S_2 + 2x} - \frac{1}{S_1 + 2x} \right) dx$$

$$= \frac{1}{2K(S_1 - S_2)} \left[\ln(S_2 + 2x) - \ln(S_1 + 2x) \right]_0^L$$

$$= \frac{1}{2K(S_1 - S_2)} \left[\ln \left(\frac{S_2 + 2x}{S_1 + 2x} \right) \right]_0^L$$

and

$$\Theta = \frac{1}{2K(S_1 - S_2)} \ln \left(\frac{S_1}{S_2} \frac{S_2 + 2L}{S_1 + 2L} \right)$$

Circular-Shaped Heat Source

In this case a circular isothermal heat source with diameter D is mounted on a substrate with thickness L as shown in Figure 10.9. The substrate is mounted on a heat sink. The thermal resistance between the heat source and the heat sink is given by

$$\Theta = \frac{1}{K} \int_0^L \frac{dx}{A} = \frac{1}{K} \int_0^L \frac{4 dx}{\pi (D + 2x)^2}$$

$$= \frac{1}{\pi K} \left(\frac{-2}{D + 2x} \right)_0^L$$

Thermal Considerations

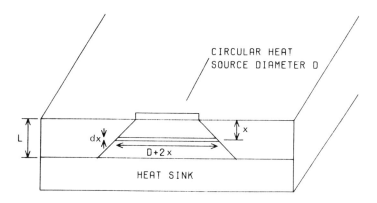

FIGURE 10.9. Circular isothermal heat source.

$$= \frac{-2}{\pi K}\left(\frac{1}{D + 2L} - \frac{1}{D}\right)$$

$$= \frac{4L}{K\pi(D^2 + 2LD)}$$

It should be noted that the calculations described above are only approximate, and care should be taken in their use. They give good "ball park" figures, however, and can prove very useful. In marginal cases it is suggested that complete computer simulation of the thermal assembly is made, and the results verified on the prototype circuits. Many computer simulation programs are available for complex thermal studies. Temperature measurements on actual circuits can be made relatively easily using an infrared microscope.

10.5 THERMAL CONDUCTIVITY OF MATERIALS

Units

Thermal conductivity of materials is specified in a wide variety of units, and some care needs to be taken when using manufacturers' data. The most common units used are shown in Table 10.1.

For today's hybrid calculations the units most commonly encountered are watts for power, °C for temperature, centimeters for dimensions, and W/cm-°C for thermal conductivity. The relationship between the units is as follows:

```
1 W   = 1 J/sec
1 cal = 4.184 J
1 Btu = 1.05435 × 10³ J
1°C   = 1.8°F
```

TABLE 10.1. Thermal Conductivity Units

Power	Temperature	Dimensions	Thermal conductivity
W	°C	cm	W/cm-°C
W	°C	in.	W/in.-°C
cal/sec	°C	cm	cal/sec-cm-°C
Btu/hr	°F	ft	Btu/hr-ft-°F

Materials

Table 10.2 gives the thermal conductivity (W/cm-°C) for the materials most commonly encountered in hybrid circuitry.

10.6 FILM RESISTOR POWER DENSITIES

Manufacturers of film pastes and depositions usually refer to a maximum power density for their pastes. This is usually about 50 W/in.^{2}. This figure is somewhat misleading in that resistor temperature is the important parameter. To avoid resistor drift and failure the film resistor should not exceed about 100 to 150°C. In an actual circuit, resistor temperature will depend on the following parameters:

TABLE 10.2. Thermal Conductivity of Materials

Material	Thermal conductivity (W/cm-°C)
Alumina (96%)	0.35
Alumina (99%)	0.37
Beryllia (BeO)	2.59
Copper	3.78
Epoxy	0.0016
Glass (lead borosilicate)	0.01
Gold	2.95
Kovar	0.19
Silicon	0.84
Silver	4.17
Solder (60-40)	0.36

1. Resistor geometries
2. Total circuit power dissipation
3. Closeness to other dissipative elements
4. Type of substrate material
5. Heat sinking used

Without considering all these factors the power density figure of 50 W/in. may be meaningless and it can be pessimistic or optimistic, depending on the operating conditions.

The results shown in Table 10.3 are a computer simulation of a resistor placed in the center of a 1 × 1 × 0.25 in. substrate as shown in Figure 10.10. The results are for alumina and beryllia substrates, with and without heat sinks. In each case the power density in the resistor was kept at 50 W/in.^{2}.

As can be seen from the table the resistor temperature is very geometry dependent. For example, a small resistor 10 × 10 mil has only a 0.42°C temperature rise on alumina at 50 W/in.^{2}, whereas a 0.2 × 0.2 in. resistor has a 120°C rise for the same power density. The equivalent temperatures when the substrate is connected to a heat sink are 0.16 and 11°C.

TABLE 10.3. Resistor at 50 W/in.^{2} in Center of Substrate

(a) Substrate in air

L (in.)	Power (W)	ΔT hot spot to ambient air (°C)	
		Alumina	Beryllia
0.01	0.005	0.42	0.28
0.05	0.125	9.16	6.70
0.10	0.50	32.81	26.23
0.15	1.125	69.97	58.41
0.20	2.000	119.88	103.14

(b) Substrate on heat sink

L (in.)	Power (W)	ΔT hot spot to heat sink (°C)	
		Alumina	Beryllia
0.01	0.005	0.16	0.03
0.05	0.125	2.22	0.48
0.10	0.50	5.18	1.31
0.15	1.125	8.25	2.37
0.20	2.000	11.00	2.48

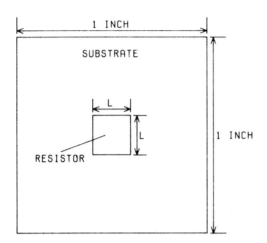

FIGURE 10.10. Film resistor power density.

In this case the 10 × 10 mil resistor on alumina substrate with a perfect heat sink would need a power density of 31,250 W/in.^{2} to obtain 100°C rise. In this case 50 W/in.^{2} would be very conservative indeed!

This example emphasizes the need for complete thermal analysis of everything but the simplest circuits. "Rules of thumb" and "intuitive guesses" can often give very erroneous results.

10.7 FACTORS FOR GOOD THERMAL DESIGN

The following guidelines should be followed for good thermal design of hybrids:

Substrate Selection

For high power substrates, beryllia is an excellent selection, as it has about six times the thermal conductivity of alumina. Alumina substrates are reasonable thermal conductors, whereas glass is a very poor thermal conductor. Glass is about 30 times worse than alumina, and glass substrates are only used in low-power applications.

Other Factors

1. When designing resistors, consideration should be given to the area reduction due to laser trimming.

2. Dissipative components should be spread evenly over the substrate area. Wherever possible, power devices or power resistors should not be placed close together.
3. The thermal paths of power components should be minimized and high thermal conductivity materials should be used to attach these components to the substrate (e.g., solder or eutectic attach should be used in preference to epoxies).
4. When using leads to conduct heat away from the substrate, high-thermal-conductivity terminals should be used (e.g., use copper instead of Kovar). Also, short, large cross-sectional-area terminals should be utilized.

11
CIRCUIT PARTITIONING

11.1 PARTITIONING

Good hybrid design should begin in the instrument or system planning phase. Sufficient time should be allocated in the beginning of the program for the system design engineers to outline their circuit requirements and to discuss them with the hybrid circuit engineer to arrive at the best approach to be implemented. With performance, size, weight, and cost in mind, the hybrid and system engineers should be able to partition the circuits to be produced into integrated circuits, thick- or thin-film hybrids, and discrete etched circuit board designs. Some of the more complex circuits may need to be partitioned into several hybrid modules, each subcircuit being selected to form as complete a circuit function as possible. This will usually simplify testing of the subcircuits and reduce overall system costs.

The partitioned subcircuits may already exist as etched circuit board modules and may need to be hybridized, for size or performance considerations. Alternatively, a subcircuit may be in the form of black box specifications, which means that a complete circuit design is needed. In either case the process of designing the prototype hybrid is fairly expensive, and it is essential that a thorough theoretical design investigation be done prior to hybridizing. Usually, this is done in the form of computer-aided-design investigations, including dc, ac, and often transient analysis of the network. Tolerances specified on components should be investigated at this point to make sure that in worst-case conditions the parts meet specifications. Tolerances should be kept as wide as possible to increase yields and as a result reduce costs. The fact that ratio tolerances are easier to meet than absolute tolerances on hybrid circuits should be kept in mind.

Since the hybrid designer is involved at this early phase of the system design, some degree of standardization of substrate sizes and packaging schemes may be possible at this time. This will again result in higher volumes of individual parts and subparts and eventually lead to lower costs.

Once the circuit design is complete, some preliminary layout and packaging schemes should be investigated. Theoretical thermal and reliability studies should be done with cost considerations in mind. Reliability criteria should match the application in mind. Tight reliability specifications required for a military project can often be relaxed considerably for a commercial application. Yield considerations are also important at this point in the design. It may be better to partition the design into two hybrids with, say, 80% yield with no rework cycle than to design one hybrid with, say, 60% yield with rework and diagnostic requirements adding to the cost.

Usually, at this point, the hybrid designer is in a position to estimate total development cost of the hybrid portion of the project, and to estimate manufacturing costs of each hybrid.

Thus at this stage in the design cycle, the hybrid designer should have completed the following tasks:

1. Circuit schematics
2. Circuit simulations and worst-case design
3. Component tolerances specified
4. Power dissipations calculated
5. Thick or thin film selected
6. Preliminary layout and package design
7. Thermal study
8. Reliability calculations
9. Development cost
10. Component manufacturing cost estimate

Figure 11.1 shows the system partitioning and design cycle for a hybrid.

Having completed all the preliminary studies the designer is now in a position to proceed with his initial hardware design.

11.2 MONOLITHIC, HYBRID, OR DISCRETE?

The first component decision to be made in the circuit partitioning is which devices should be integrated, which should be hybrid, and which should be discrete.

Often this decision is very difficult to make since there may be conflicting variables. However, there often are some relatively easy decisions that can be made. For low-frequency circuits where size and weight have little significance and off-the-shelf components are available, it is usually quicker and costs less to use discrete circuitry on an etched circuit board. Where large quantities of small-scale integrated circuits (SSI) are being used, it may be advantageous to

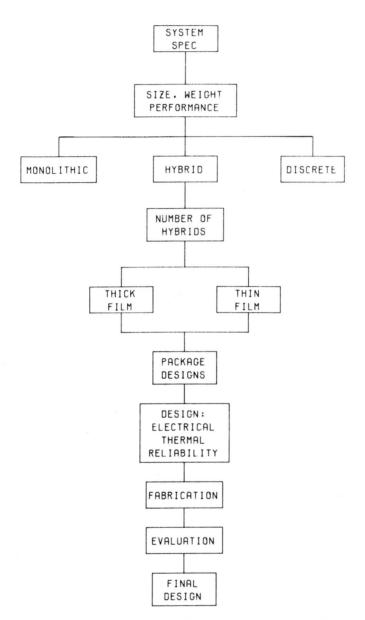

FIGURE 11.1. System partitioning and design cycle.

Circuit Partitioning 141

design a custom large-scale integrated circuit (LSI). However, the cost of developing such an LSI circuit may be very high, often greater than $100,000, and the potential volume needs to be high to justify this cost. Where weight and size are important and the component volume is low, a custom hybrid LSI circuit can be developed for much less than the monolithic circuit; costs are usually $25,000 to $50,000. However, the manufacturing cost of the hybrid will be much greater than that for the monolithic IC circuit. There will therefore be a conflict between development cost and manufacturing cost.

Precision circuitry is difficult with both discrete and integrated circuits. Some companies are laser-trimming film resistors on monolithic ICs to obtain precision circuitry. However, the process is very expensive to develop and requires a large capital investment. Precision laser trimming on hybrid circuitry is much less complex and requires less capital investment. Thus hybrids offer a good solution to circuits such as analog-to-digital and digital-to-analog converters, and precision analog circuitry.

High-frequency circuits above a few hundred megahertz are difficult to implement with discrete circuitry due to the parasitic reactances associated with ECB design. Also, many components cannot be implemented on integrated circuits (e.g., large capacitors, inductors, matching networks). However, hybrid circuitry is ideal for these high-frequency applications. Parasitic reactances can be reduced considerably due to tighter package designs, and the use of semiconductor dies eliminates the package capacitance of discrete components. Inductors and large-value capacitors are readily available for hybrids. At microwave frequencies hybrids have virtually captured the whole-component market. Using microwave transistors and diodes, a wide range of microwave circuits can be designed. Matching networks using interdigital capacitors, film inductors, and transmission lines can all be designed on the hybrid.

11.3 THICK OR THIN FILM?

Having determined which components should be hybridized, the next decision is which film technology should be used, thick or thin?

Often, only one technology is available and that technology may be utilized exclusively. However, in many cases both technologies are available either in-house of from external manufacturers.

Cost is usually one of the major decision criteria involved in any design. The substrates used for thick film do not need very critical specifications, whereas for thin-film substrate, specifications such as surface finish and camber are important. Thus thin-film substrates are usually more expensive than thick-film substrates; e.g., a 3 × 3

TABLE 11.1. Comparison of Thick and Thin Films

Parameter	Thick film	Thin film
Resistors (value)	High	Low
Resistors (stability)	Good	Excellent
Resistors (tolerance)	Low	Lowest
Capacitors (value)	High	Low
Parasitics	Low	Lowest
Frequency limits	Medium	High
Package density	Medium	High
Cost	Low	Medium

in. thin-film substrate may cost $5 compared with 50 cents for a 3 × 3 in. thick-film substrate. Also, the processing costs involved in thin film may be much higher than those for thick film. It may cost $20 to $50 to process the 3 × 3 in. thin-film substrate and only $2 to $5 for the equivalent thick-film substrate. However, some of these costs can be offset by the fact that higher component densities can be obtained with thin film. Individual circuits can be stepped-and-repeated on the substrates; e.g., an individual circuit may only be 1/2 × 1/2 in. in thin film and 36 circuits can be produced on one 3 × 3 in. plate. The equivalent circuit in thick film would occupy maybe 1 × 1 in. and only nine circuits could be achieved on a 3 × 3 in. substrate. The overall result is that a thin-film substrate may cost only two to three times the equivalent thick-film circuit. This may or may not be an

TABLE 11.2. Comparison of Thick- and Thin-Film Performance Characteristics

Parameter	Thick	Thin
Resistivity (Ω/square)	3 Ω–10 MΩ	25–100Ω
TCR (ppm/°C)	±50 to ±150	100
TCR tracking (ppm/°C)	30	5
Noise (μV/V)	3.0	0.05
Stability at rated load, 1000 hr, + 70°C ambient (%)	±1.0	0.06
Tracking stability at rated load 1000 hr, + 70°C (%)	±0.05	0.01
Voltage coefficient (%/V)	0.001	0.0005
Dynamic trim	Yes	Yes

Circuit Partitioning 143

TABLE 11.3. Applications—Thin Film Vs. Thick Film

	←	High frequency	
	←	High density	
	←	High tolerance	hermetic-sealed from environment
	←	High tolerance/low density	
	←	Time/temperature tracking	
		High volume →	
	←	Low volume	
	←	Low development time →	
Thin film	←	Low development cost →	Thick film
(photoetch)		Nonhermetic →	(screened)
	←	Plug-in substrate →	
		Second source →	
	←	Low noise	
		Breadboard →	
	←	Modify design →	
	←	Producibility →	
	←	Reliability →	
	←	Hermeticity →	

Clearly thin	Questionable	Clearly thick
>100 MHz or 0.1%	50-100 MHz or some 0.1%	<50 MHz and 0.1%
>20 components/in.^{2}	15-20 components/in.^{2}	<15 components/in.^{2}
High reliability	Medium reliability	Industrial equipment
Low volume	Medium volume	High volume
Low noise	Medium noise	Noise not critical

important factor in the overall assembled hybrid cost. Often, the processed substrate cost is only a minor fraction of the final component cost. In high volumes with automated screening and firing equipment, however, thick film can result in very low cost consumer-type components.

Performance is the second major factor in a comparison of thin and thick films. The major performance advantage of thick film over thin film is the availability of high-value resistors. The major performance advantages of thin film over thick film are higher-resolution conductors and resistors, better resistor stability, and better high-frequency performance.

In general, thick film is used in low-frequency, low-cost, high-volume products, whereas thin film is used in precision circuitry and at very high frequencies. Tables 11.1, 11.2, and 11.3 summarize the comparison between the technologies.

12
DESIGN CYCLE

12.1 DESIGN CYCLE

Whether the hybrid is designed in-house or designed by a hybrid vendor, it is important to understand the complete hybrid design cycle.

If no in-house facility is available, the hybrid manufacturer may perform the entire design function. In other cases the customer may supply the hybrid layouts and bill of materials and the hybrid manufacturer simply processes the components with little or no redesign.

The overall cycle is indicated in Figure 12.1. The design may be the hybridization of an original circuit to meet some "black box" specifications. In both cases circuit analysis and simulation are necessary to ensure that the circuit will meet its specifications in worst-case conditions.

Decisions are then made concerning the choice between thick- or thin-film technologies. Generally, high-precision circuits such as analog-to-digital and digital-to-analog converters and high-frequency circuitry will require thin film, whereas the majority of other circuitry will use thick film.

A preliminary layout will now be attempted to determine both substrate size and packaging requirements.

Process and product flowcharts can now be determined and the final layout completed. Masks and screens are processed and prototype hybrids fabricated. These circuits can be tested to meet the component specifications. Generally, one or more design cycles are required before the component meets all its specifications.

12.2 CIRCUIT ANALYSIS AND SIMULATION

In the final packaged form hybrid circuits are usually difficult to adjust. Resistors are usually in film form and cannot be changed easily. At best they can be increased in value by scribing with a diamond scribe

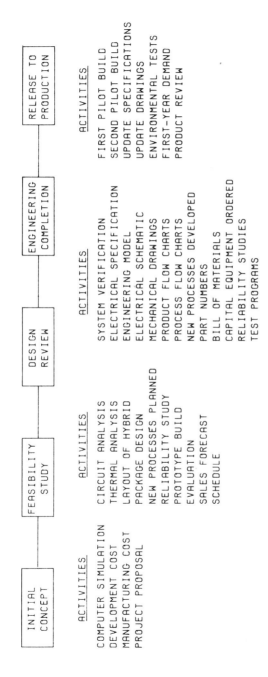

FIGURE 12.1. Overall design cycle.

or cutting with a laser beam. Bonded components such as integrated circuits and discrete components are also difficult to remove and replace. For these reasons the "cut and try" method often associated with discrete etched circuit board design is not good practice for hybrid design. It is important that resistors, capacitors, inductors, and other component values are determined fairly precisely prior to committing the design to the layout stage. For low-frequency circuitry this can be achieved easily. However, at high frequencies or with high-precision circuitry, it is essential that the circuit be analyzed thoroughly prior to layout. This will usually involve computer simulation and characterization of components used in the design.

There are a wide variety of computer simulation programs available, including linear, nonlinear, and transient analysis. Data can usually be obtained from component manufacturers on the modeling and characterization of semiconductor devices. Alternatively, internal programs can be set up to fully characterize and model the component used. For high-frequency and microwave design, network analysers are available to give accurate measurements of device scattering parameters. S-parameter analysis and optimization programs are also available commercially. Optimization programs can be extremely useful in design, as they not only analyze a circuit's performance, but can adjust selected element values to meet a particular design goal.

Following initial analysis and optimization the circuit should be analyzed for worst-case performance. Components can be changed for both tolerance and temperature effects.

Although extensive circuit analysis may seem a time-consuming and expensive exercise, it may prove invaluable in the long run. Without worst-case analysis, for example, a circuit may meet some specifications for several months into production until a worst-case lot of devices is encountered and yield may suddenly drop to almost zero. Redesign of the circuit when it is well into production may prove an expensive and embarrassing mistake.

Once you are satisfied that the circuit will meet its design goals, the design can proceed to the layout stage.

12.3 LAYOUT

The initial layout will be a "single line" drawing, to establish approximate substrate and package pin configuration. Early decisions on package pin configurations may be useful to the system designer, who may be designing circuits to interface with the hybrid. EC board design can then proceed in parallel with the hybrid design.

The hybrid designer will now need to list device sizes and any critical areas of layout (e.g., crosstalk considerations, parasitic elements, etc.). With these constraints in mind an accurate layout is done on gridded paper at a 10:1 or 20:1 scale. The layout person should be aware of all the constraints of thick- and thin-film guidelines. Resistors are designed for the required power dissipation, voltage stress, and trim ratios. The various layers are usually identified by different colors. Once the layout is complete it should be checked thoroughly and if high-frequency considerations are a potential problem, stray reactance should be calculated and the circuit again simulated on the computer.

Once the layout is complete, photographic masks can be generated. One common way to do this is to cut Rubylith artwork from the layout drawing. There will be one "Ruby" per layer. The Rubylith masters are reduced on photographic film. The problem with this approach is that design changes may involve recutting a new set of artwork. An alternative approach is to digitize the artwork on a computer controlled digitizer, store the X-Y coordinates of each layer on magnetic tape, and use the tape on a photographic pattern generator. Using this method design changes are easier to implement.

The photographic masks generated are used to produce screens for thick film or used directly for the photoprocessing of thin films.

12.4 PROCESSING AND PACKAGING

The substrates can now be processed in the appropriate thick- and thin-film departments. Usually, for initial evaluation, runs will be from 10 to 50 parts. The substrates then proceed to laser trim and finally packaging. There should be good communications at this point back to the hybrid designer concerning processing of the part. Even the best designs following all the known rules will have some processing problems.

Typical problems are screening and alignment difficulties, laser trim difficulties such as probing and trim times, and also packaging problems such as die attach and bonding difficulties. The designer should be responsive to these problems and attempt to alleviate them in redesign. Yield problems encountered in the packaging area may mean complete redesign including device changes and even repartitioning of the circuitry.

12.5 TEST AND REDESIGN

The final stage is testing and evaluation in the system. The hybrid should be thoroughly evaluated as an individual component and a specification written around it.

The device may be capable of a wider specification than that dictated by the system it was designed for. Thus, if a complete evaluation is made of the device and the results cataloged, further applications may be found for the hybrid. This may result in higher volumes and often lower manufacturing costs. Also, the development costs will be spread over a larger volume of parts.

Tests may indicate some design changes, and these should be made together with those dictated by processing and package considerations. Most designs should be completed with one or two redesigns.

12.6 RELIABILITY AND LIFE TESTS

A final reliability evaluation should be performed after all the steps in the design cycle have been completed and the design is frozen and ready for production. Typical reliability and life tests are described below.

High-Temperature Storage (Nonoperating)

This test is designed to check for (1) electrical parameter drift, (2) corrosion, (3) defects on silicon die, and (4) metallization defects. It is intended to accelerate drift with time and temperature. Devices are electrically tested at various intervals during the test. Typically, the test requires storage in a prescribed high ambient temperature of about 150°C for up to 1000 hr.

Temperature Cycling

This is an essential test. It is inexpensive and effective and it checks for (1) packaging and sealing defects, (2) metallization and lead defects, and (3) theoretical mismatch between various materials used in the hybrid.

An example of such a test is 10 temperature cycles from -65 to +200°C with a minimum of 15 min at each extreme and a maximum of 2 min at room temperature between extremes.

Thermal Shock

The test checks for (1) package and seal defects, (2) bonding and metallization defects, and (3) thermal mismatch.

Although this test is similar to temperature cycling, there is a significant difference: the severe strain generated by the extreme temperature difference is not indicative of any actual operation conditions. However, it is more useful for indicating possible problem areas.

An example of a thermal shock test is immersion in a liquid at 125°C for not less than 5 min, and immediate transfer to a liquid bath at -65°C for not less than 5 min. This is repeated for five cycles.

Design Cycle 149

Accelerated Life Tests

Accelerated life tests are designed to accelerate the failure mechanisms of a circuit in order to reduce the number of samples and test time required to obtain useful reliability data. Usually, the acceleration factor is a thermal or electrical stress greater than that expected for the circuit in normal operation.

Accelerated life tests are used for the following:

1. To predict failure rates at lower stresses
2. To compare relative failure rates
3. To assess potential problem areas

Using accelerated tests to predict failures, however, opens the door to many uncertainties regarding the validity of predicted failure rates. Any errors can be minimized if the results of an accelerated test on a new circuit can be compared with those on a circuit with a known reliability level.

Other Tests

Many other tests are available for testing the device for possible failure mechanisms (e.g., moisture-resistance tests, corrosion test in a sulfide atmosphere, shock and vibration tests, etc.).

Usually, the tests used depend to a great extent on the type of circuit, package, and so on, and will vary from design to design.

12.7 TESTING OF HYBRID CIRCUITS

An important and often overlooked part of the design cycle is the program for the final electrical testing of the completed hybrid. Usually, because of the complexity and size of the hybrid it is not convenient to monitor test points as may be done on discrete circuitry. Usually, the hybrid is powered and its input/output characteristics measured. Typical measurements will include bandwidth, rise times, and aberrations for analog circuitry, logical functions for digital circuitry, and two-port S parameters for microwave circuitry.

If the hybrid is complex and expensive to manufacture, some rework is usually required on faulty hybrids. This rework cycle will improve yield, resulting in lower costs. For efficient rework cycles the designer needs to develop a diagnostic package in order to indicate which die or wire bond on the circuit is faulty. Appropriate test pads are often added to the substrate to facilitate this fault diagnosis. Designing with this testability and diagnostic capability in mind will usually put some constraints on the layout and electrical design but can give significant rewards in the long run.

Pretesting of components is also a very important consideration. If the dice are known to be functionally well before being bonded to the hybrid, there is a higher probability of a good hybrid yield. Full functional testing of semiconductor dice is not always possible. However, the die can be prepackaged in chip carriers or similar packages, and can be tested and often burned in prior to assembly on the hybrid. This improves the overall yield at the expense of additional cost due to the chip carriers. The designer needs to examine the trade-offs between increased yield and higher material cost.

The hybrid designer should also be involved in the design of the test fixtures. Hybrids that are stable in their operating environment may well oscillate in a poorly designed test fixture. Decoupling of power supplies is usually important, and parasitic reactances due to test probes and connections should be minimized.

12.8 OVERALL DESIGN CYCLE

The typical steps in the design cycle for a hybrid from initial concept through production are shown in Figure 12.1. Typical milestones and development times are indicated.

Feasibility Milestones

This usually occurs about 2 to 4 weeks after initial discussion between the hybrid and the system designer, and after initial study by the hybrid designer as to the feasibility of the product. The initial study may just be on paper and will often involve some computer simulation, and estimates of development and manufacturing costs. Also, considerations such as thick or thin film, type of package, test procedures, and reliability will be involved. Usually, a proposal will be written and presented to the user group for approval. Once approved, layout is started and prototypes are built.

Hybrid Design Review Milestones

This will occur at the completion of the first build of breadboard parts and evaluation of the hybrid in the system. This usually occurs about 8 to 10 weeks from initial concept. Usually, the following data are available at this milestone:

1. Circuit analysis/synthesis results
2. Tentative electrical specifications
3. Thermal analysis data
4. Reliability analysis data
5. Data on evaluation in the system

Design Cycle 151

There is usually a thorough review of both the electrical and mechanical characteristics at this time. This should be the hybrid "go" or "no go" decision point.

Hybrid Engineering Completion Milestone

This will occur after the first engineering parts have been built and evaluated in the system. This is usually about 8 to 10 weeks from the design review. The following data should be available at this milestone:

1. Firm electrical specifications
2. Engineering schematics, including substrate drawings, layout drawings, bonding diagram, package outline, and assembly drawings
3. Test specifications
4. Subpart specifications
5. Bill of materials
6. Capital equipment requirements
7. Reliability information

Manufacturing should be involved by this milestone and should have had a "buildability" meeting to discuss whether new processes or equipment may be needed in manufacturing.

Hybrid Release to Production Milestone

Usually, this will follow two pilot builds of the parts. The first build is usually in engineering with manufacturing being trained, and the second build is usually in manufacturing with help from engineering. During these two builds all the problems should be identified and solved. Also, all the assembly drawings and specifications should be updated. During this period from engineering completion to production release, some environmental testing of parts should also be completed.

If the design has been thorough with sufficient consideration given to electrical, thermal, packaging, and reliability problems, we should have a smooth transfer to production.

13
RELIABILITY

13.1 RELIABILITY ASPECTS OF HYBRIDS

No component will last forever; even the best designed and assembled component will eventually fail. The object of reliability considerations is to prevent a high incidence of failures during the normal expected lifetime of a component. All aspects of design and manufacturing should be considered in order to improve the reliability of a component.

Hybrids utilize similar processes to those used in conventional integrated circuit and etched circuit board assembly: die attach, wire bonding, soldering, and sealing. There is therefore no fundamental reason why hybrids should be any less reliable than the equivalent discrete circuit. In fact, because of the better thermal characteristics of hybrids and usually a reduction in the total number of connections, hybrids should have better reliability than the equivalent discrete circuit.

13.2 RELIABILITY DATA

Reliability considerations should be as important a part of the design cycle as electrical, mechanical, and thermal considerations. The designer needs to consider the following questions:

1. What is the expected lifetime of the hybrid?
2. What is the random failure rate of the hybrid?
3. What failure modes are likely to be encountered?
4. What is the probable result of both mechanical screening and electrical burn-in on the hybrids?
5. What types of climatic and mechanical environments will the component encounter?
6. Will accelerated stress testing give meaningful results as to predicting the hybrid's reliability?

13.3 RELIABILITY DEFINITIONS

Environment. The physical conditions which a component may be exposed to during storage or operation. Environment usually covers climatic, mechanical, and electrical conditions.

Reliability. The probability of the component performing within specifications for a given period of time in a specific environment. The reliability of a component is highly dependent on the specified environment.

Failure. A part that no longer meets its performance criteria. Failures include devices that have drastically failed as well as components that function but are out of specification. For example, an amplifier specified with a gain of 10 ± 1 dB is termed a failure if it is open circuit (0 dB) or out of specification (i.e., less than 9 dB or greater than 11 dB).

Failure Distribution. The distribution of failures plotted as a function of time. This is usually plotted for a particular group of parts operating in a particular environment.

Infant Failures. These are failures that occur early in the life of a group of parts. Such failures are usually associated with production problems not normally encountered (e.g., mechanical failures due to poor die attach or wire bonds, missed inspection problems, or contamination). Infant failures are usually characterized by a decreasing failure rate with time.

Failure Rate. This is the number of failures per unit time. The failure rate is usually expressed as %/1000 hr.

Constant Failure Rate. After infant failures have been removed from a group of parts, failures that occur in a completely random fashion will result in a constant failure rate. If the events are random, one failure does not influence the probability of future failures.

Exponential Failure Distribution. This is the failure distribution of a group of parts that have a constant failure rate. After one fails, the probability is the same that the remaining parts will survive the same length of time. The exponential curve results because of the diminishing quantity remaining in the given group of parts. Figure 13.1 shows an exponential failure distribution.

Wear-Out Failure. This is a failure caused by a mechanism that is related to the physics of a device, its design, and process parameters. Wear-out failures should be distinguished from random failures, which

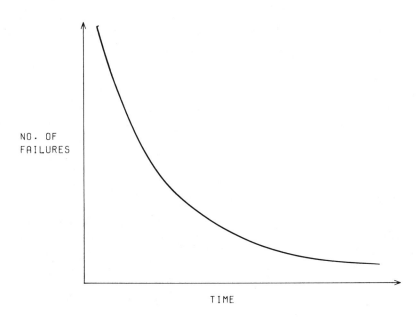

FIGURE 13.1. Exponential failure-rate distribution.

are associated with the variability of workmanship quality. The component should be designed such that, if manufactured and used within specifications, the wear-out failures occur well beyond the intended use period. Examples of wear-out failure mechanisms are metal migration, resistor drift, wire bond fatigue failures, and IC failures. Reliability life testing should include evaluation of such wear-out modes.

Acceleration Factor. The major failure mechanisms of a component stem from electrical aging and both electrical and mechanical wear.

Electrical and mechanical wear will include such mechanisms as temperature cycling; shock and vibration; variations in power supplies, including switching transients; and corrosion effects.

Electrical aging of semiconductor, capacitor, and resistor components can be accelerated at elevated temperatures. The electrical aging is a chemical process generally following the chemical reaction equation of Arrhenius:

$$F = Ae^{\frac{-E_a}{kT}}$$

where

F = failure rate
A = a constant

E_a = activation energy, eV
k = Boltzmann's constant ($8.6_{10}{}^{-5}$ eV/K)
T = absolute temperature, K

Since electrical aging is accelerated at increased temperature, we can define a time acceleration factor:

$$\text{Time acceleration factor} = \exp\left[\frac{E_a}{k}\left(\frac{1}{T_1} - \frac{1}{T_2}\right)\right]$$

where

T_1 = reference temperature, K
T_2 = acceleration temperature, K

Figure 13.2 shows acceleration factor as a function of temperature for an activation energy of 1 eV. This acceleration factor is used to speed up reliability estimates of failure rates. The failure rate is measured with the component operated at an elevated temperature such as 100°C. The acceleration factor is then used to calculate the failure rate at normal operating temperatures. Care should be taken when choosing the elevated temperature, since high temperatures may introduce additional failure mechanisms and occasionally prevent failure mechanisms such as humidity problems.

The introduction of new failure mechanisms will result in convervative reliability estimates, whereas prevention of failures may result in overoptimistic reliability failures, which can be a costly error.

For typical temperatures between room temperature (293 K = 20°C) and 100°C = 373 K it is a good approximation that the acceleration factor doubles for every 10°C rise in temperature.

An important constant in the acceleration factor is the activation energy, E_a. This constant ranges from 0.7 to 1.3 eV, with 1 eV being a widely accepted value that is often used in place of more exact data.

13.4 CONSTANT FAILURE RATE

After infant failures have been removed and before wearout failures begin to occur, most components exhibit a constant failure rate. The typical failure rate curve for a component is shown in Figure 13.3.

The infant failures should be eliminated by an appropriate burn-in program, and wearout failures should occur well past the useful life of the component.

The components that are "shipped" should thus have a constant failure rate. The failure distribution for any given lot of parts will therefore be exponential. As shown in Figure 13.1, after one fails the probability is the same that the remaining parts will survive the same length of time. The exponential curve results because of the diminishing quantity remaining in the lot.

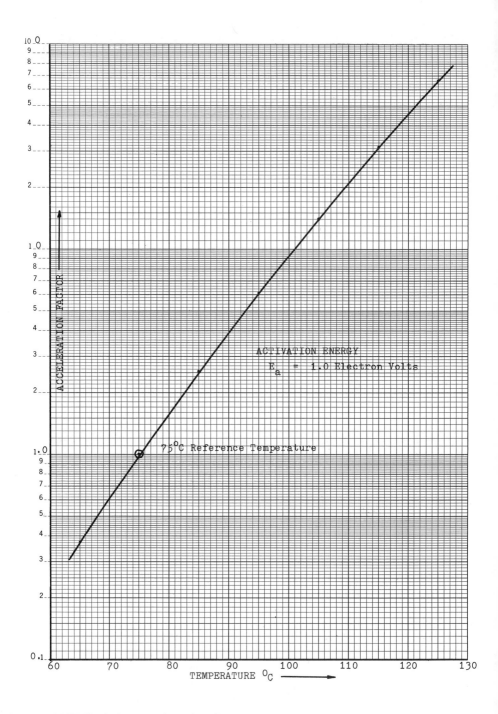

FIGURE 13.2. Acceleration factor vs. temperature.

Reliability 157

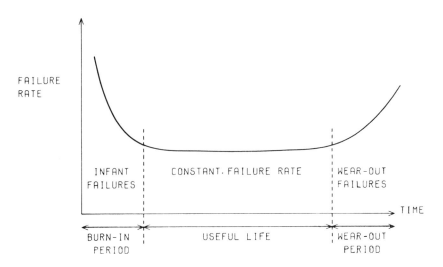

FIGURE 13.3. Typical "bathtub" failure curve for electrical components.

The probability of a component surviving for a time t is R, where

$$R = e\left(\frac{-t}{MTTF}\right)$$

and mean time to failure

$$MTTF = \frac{1}{\text{failure rate}}$$

For any given group of parts,

$$MTTF = \frac{\text{total unit operating hours}}{\text{number of failed parts}}$$

The term "mean time between failure" (MTBF) is often used instead of "mean time to failure" (MTTF). However, MTBF should be applied only to systems that are repairable. In the case of hybrids and most other components, the systems are usually unrepairable and the term "mean time to failure" is more appropriate.

Although MTTF may be a large number of hours, it should be noted that when the mission time is equal to the MTTF, the probability is 0.37. This means that only 37% of a population is surviving by the time MTTF is reached. In a similar manner, at 10% MTTF, 90% survive; at 1% MTTF, 99% survive. For example, with an MTTF of 100,000 hr and constant failure rate, 10% failures will occur in 10,000 hr.

13.5 FAILURE RATE OF MULTIPLE-ELEMENT SYSTEMS

A hybrid is built from a variety of components and assembly processes, each of which has its own failure rate. For example, wire bonds have a failure rate, die attach has a failure rate, as does a semiconductor die or chip capacitor. One would expect a higher hybrid failure rate than each individual element failure rate.

The failure rate L_t of such a multielement system is given by

$$L_t = L_1 + L_2 + L_3 + \cdots + L_n$$

This assumes constant failure rate in each element of the system. The MTTF of the hybrid will be

$$\text{MTTF} = \frac{1}{L_t}$$

Normally, the failure rate is known for most elements in the system. For example, a wire bond may have a failure rate L_{wb}. If there are N wire bonds in the circuit, wire bonding will contribute NL_{wb} to the system failure rate. Fortunately, the failure rates for wire bonding are very low, usually about 0.0001%/1000 hr. Even a complex hybrid with 200 wire bonds will only have a failure-rate contribution of 0.02% per thousand hours due to wire bonding.

Typical failure rates for elements in a hybrid circuit are given in Table 13.1. Care should be taken when using these figures since they are very process dependent. Wherever possible, numbers should always be generated for one's own processes. However, these numbers can be used to generate ballpark numbers for estimating failure rates. The actual failure rate of any component should be proven by extensive testing of that component. Any theoretical estimate is usually used only for circuit partitioning and cost estimates.

13.6 ESTABLISHING RELIABILITY GOALS

The major question facing the hybrid designer in determining reliability goals is: What does the system need and what will it cost? Compromises can be made with reliability as with any other performance parameter. However, reliability is based on statistical data and is much less predictable or measurable than most other parameters. Because of this lack of predictability, accuracies of two-, five-, and even tenfold can be expected. Although this may seem extremely inaccurate, we are usually talking about making the product better than some lower limit, knowing that often some units could be easily tenfold better.

A mathematical model based on the failure rates of the elements of the system will yield a ballpark figure that is a good starting point. The model should be tuned to reflect test and field experience and can become a very useful tool in avoiding dismal field reliability.

TABLE 13.1. Hybrid Circuit Element Base Failure Rates (%/1000 hr)

Element	Temperature (°C)				
	25	50	75	100	125
Thick-film resistor	0.0005	0.0010	0.0015	0.002	0.0025
Chip capacitor	0.001	0.0015	0.0025	0.006	0.025
Wire bonds					
Au-Al ball	0.000005	0.00002	0.0001	0.001	0.006
Al-Au	0.00001	0.00001	0.00001	0.00001	0.00005
Al-Al	0.00001	0.00001	0.00001	0.00001	0.00001
Au-Au	0.000004	0.000004	0.000004	0.000004	0.000004
Crossovers	0.000005	0.000005	0.000006	0.000008	0.00001
Transistor chips					
Low power	0.0001	0.0003	0.0009	0.0027	0.007
Power	0.005	0.010	0.03	0.09	0.27
Diode chips	0.0001	0.0003	0.0009	0.0027	0.007
Microcircuits					
Quad gate or equivalent	0.002	0.0036	0.018	0.082	0.24
Dual flip-flop or op amp equivalent	0.004	0.0072	0.036	0.164	0.48
SSI (equivalent of 25 gates)	0.0125	0.0225	0.1125	0.512	1.5
MSI (equivalent of 50 gates)	0.025	0.0459	0.225	1.02	3.0
LSI (equivalent of 100 gates)	0.050	0.09	0.45	2.04	6.0

Other failures, such as learning factors and environmental factors, can be added in to refine this model. For example, the learning factor will account for the fact that manufacturing is on a learning curve with any new product, and product quality and reliability usually improves along this learning curve. Environmental factors will modify the reliability figures depending on the type of environment the component is expected to encounter. Obviously, how one uses these "fudge" factors will drastically influence the final estimates. However, with experience and correlation with test and field results, reasonably accurate models can be generated.

13.7 RELIABILITY TESTING

The reliability goals of the hybrid component should be verified prior to release to manufacturing. A given group of parts are usually tested under accelerated conditions to determine if they meet the reliability goals. For the results to be valid, the accelerated conditions must not alter the basic modes and mechanisms of failure.

Usually, a group of 50 to 100 parts are tested at some chosen high operating temperature, usually 100 to 150°C. The parts are pretested and then operated for a fixed period of time such as 1000 hr. The parts are then retested and the number of failures recorded. Assuming that the failures are random, the MTTF is given by

$$\text{MTTF} = \frac{\text{total device hours}}{\text{number of failures}} \times \text{acceleration factor}$$

where the acceleration factor is given by

$$\text{Acceleration factor} = \exp\left[\frac{E_a}{k}\left(\frac{1}{T_{normal}} - \frac{1}{T_{acceleration}}\right)\right]$$

where

E_a = activation energy (~ 1 eV)
k = Boltzmann's constant (8.6_{10}^{-5} eV/K)
T_{normal} = normal expected operating temperature, K
$T_{acceleration}$ = acceleration temperature, K

Example Suppose that a group of 100 parts were operated at an accelerated temperature of 150°C for 1000 hr with two failures. The maximum operating temperature is 75°C. The resulting MTTF is

$$\text{MTTF} = \frac{100 \times 1000}{2} \; e^{(1/8.6_{10}^{-5})(1/348 - 1/423)}$$

$$= 1.87_{10}7 \text{ hr}$$

Reliability

and

$$\text{failure rate} = \frac{1}{\text{MTTF}} = 0.005\%/1000 \text{ hr}$$

Confidence in Measured MTTF

The mean time to failure calculated from the accelerated test data is only an estimate. The degree of precision or confidence we have in the answer also needs to be considered. Usually, only a small number of parts are tested for a relative small period of time, resulting in small numbers of failures. For such small numbers of failures this can lead to misleading estimates of MTTF. In the example above, zero failures would lead us to believe that the MTTF is infinity, whereas one failure leads us to an estimate of MTTF = $3.7_{10}7$, two failures leads us to an MTTF = $1.87_{10}7$. Obviously, in this case the MTTF estimate is extremely sensitive to the number of failures. One way out of this is to run the test for much longer periods of time, to give a higher confidence in the result. However, usually this is impractical. Even 1000 hr is approximately 6 weeks, a fairly long time compared with the overall component development time.

One way out of this is to use statistical methods to improve the confidence level in the MTTF estimate. A chi-square distribution is applied which describes the probability of the MTTF number representing the true distribution of the population. A lower confidence limit can be specified (e.g., a 90% lower confidence limit implies that 9 times out of 10 the MTTF number will be greater than the one calculated). The result is that the MTTF estimate from the test data is divided by a statistical factor instead of the measured number of failures.

$$\text{MTTF} = \frac{\text{total device hours}}{\text{statistical factor}} \times \text{acceleration factor}$$

For a test that is stopped at a given time (not at a given number of failures) and for a 90% lower confidence limit, the statistical factor is given in Table 13.3.

Note that as the failures increase, the ratio between the statistical factor and the number of failures become smaller. When the failures are very large, the statistical factor will equal the number of failures. However, for practical reasons, tests are seldom run for that length of time.

Using the data from the previous example we see that two failures corresponds to a statistical factor of 5.3:

$$\text{MTTF} = \frac{3.74_{10}7}{5.3} = 7_{10}6 \text{ hr}$$

This figure has a 90% lower confidence limit.

Similar statistical factors can be obtained for 95%, 99%, or any other lower confidence limit. However, in most cases the 90% lower confidence limit is sufficient.

13.8 FAILURE MECHANISMS IN HYBRIDS

A major cause of failures in electronic systems has been failures in component interconnections, particularly solder interconnects. Using hybrid circuits the number of manually made component interconnects is reduced dramatically. Some of these interconnects are replaced by chemically bonded material interfaces on the substrate, the film components, such as resistors, conductors, and capacitors. This reduces the module's susceptibility to wiring errors, and damage due to dynamic environments, such as shock, vibration, and acceleration.

Localized heating and hot spots within resistive elements are reduced due to the direct bond between the films and the usually good thermally conductive substrate. This results in very reliable resistive films. However, the films will drift with time, typically 0.25% for thick film and 0.1% for thin film. Such drifts should be allowed for in any worst-case analysis.

The major failure mechanisms in the hybrid are in the add-on components, such as chip capacitors, chip resistors, transistors, diodes, integrated circuits, and wire interconnects (wire bonds). Table 13.2 shows comparative failure rates for various wire-bonding techniques, and Table 13.1 shows typical element failure rates.

Although a single wire bond is very reliable, there may be more than 200 wire bonds on a complex hybrid, and they may make a major contribution to the failure rate.

In partitioning it is important to include these reliability figures in the design stage. It may be necessary to partition into several small hybrids rather than one large hybrid, to prevent one failure mechanism from dominating the failure-rate calculation.

TABLE 13.2. Comparative Failure Rates for Various Bonding Techniques (%/1000 hr)

Interconnection	One lead	14-lead device	150-lead device
Thermocompression wire bonds	0.00013	0.0018	0.02
Untrasonic wire bonds	0.00007	0.001	0.014
Face bond	0.00001	0.00014	0.0015
Beam lead	0.00001	0.00014	0.0015

TABLE 13.3. Statistical Factors for a 90% Lower Confidence Limit

Failures measured	Statistical factor
0	2.3
1	3.9
2	5.3
3	6.7
4	8.0

13.9 FAILURE ANALYSIS

Detailed knowledge of hybrid microcircuit failure modes and failure mechanisms is the key to producing reliable devices. A thorough understanding of the causes of failure allows meaningful improvements in the areas that will increase component reliability.

The main steps in a failure analysis procedure are to determine the following:

1. Effect of the failure
2. Failure mode
3. Characterization of abnormalities associated with the failure mode
4. *Failure mechanism hypothesis*: Relation of possible stresses to the abnormalities determined in step 3
5. *Verification*: Carry out tests to verify the failure mode and cause
6. *Corrective action*: Apply action to the design or process to rectify failure mode

Such failure analysis procedures are essential to a successful hybrid operation.

13.10 OTHER RELIABILITY TESTS

Besides acceleration testing of the component to estimate failure rate, other routine tests are performed on the hybrid. Some of these are described in this section.

High Temperature Storage (Nonoperating)

This test is designed to check for (1) electrical parameter drift, (2) corrosion, (3) defects on silicon die, and (4) metallization defects. It is intended to accelerate drift with time and temperature. Devices are electrically tested at various intervals during the test. Typically, the test requires storage in a prescribed high ambient temperature of about 150°C for up to 1000 hr.

Temperature Cycling

This is an essential test. It is inexpensive and effective and it checks for (1) packaging and seal defects, (2) metallization and lead defects, and (3) thermal mismatch between various materials used in the hybrid.

An example of such a test is 10 temperature cycles from -65 to +200°C with a minimum of 15 min at each extreme and a maximum of 2 min at room temperature between extremes.

Thermal Shock

The test checks for (1) package and seal defects, (2) bonding and metallization defects, and (3) thermal mismatch.

Although this test is similar to temperature cycling, there is a significant difference: the severe strain generated by the extreme temperature difference is not indicative of any actual operating conditions. However, it is more useful for indicating problem areas.

An example of a thermal shock test is immersion in a liquid at 125°C for not less than 5 min, and immediate transfer to a liquid bath at -65°C for not less than 5 min. This is repeated for five cycles.

Other Tests

Many other tests are available for testing the device for possible failure mechanisms (e.g., moisture-resistance tests, corrosion test in a sulfide atmosphere, shock and vibration tests, etc.).

Usually, the tests used depend to a great extent on the type of circuit, package, and so on, and will vary from design to design.

GLOSSARY OF TERMS

Abrasive trimming Trimming a film resistor to its nominal value by notching the resistor with a finely adjusted stream of an abrasive material, such as aluminum oxide, directly against the resistor surface.

Active components Electronic components, such as transistors, diodes, thyristors, etc., which can operate on an applied electrical signal so as to change its basic character; i.e., rectification, amplification, switching, etc.

Active devices Discrete devices such as diodes or transistors; or integrated devices, such as analog or digital circuits in monolithic or hybrid form.

Active element An element of a circuit in which an electrical input signal is converted into an output signal by the nonlinear voltage/current relationships of a semiconduct device (*see* Active components).

Active network A network containing active and passive elements.

Active substrate A substrate in which active and passive circuit elements may be formed to provide discrete or integrated devices.

Active trim Trimming of a circuit element (usually resistors) in a circuit that is electrically activated and operating to obtain a specified functional output for the circuit (*see* Functional trimming).

Add-on component Discrete or integrated prepackaged or chip components that are attached to a film circuit to complete the circuit functions.

Add-on device Same as Add-on component.

Adhesion The property of one material to remain attached to another; a measure of the bonding strength of the interface between film deposit and the surface which receives the deposit; the surface receiving the deposit may be another film or substrate.

Alloy A solid-state solution of two or more metals. (v.) To melt or make an alloy.

Printed with permission of the International Society for Hybrid Microelectronics.

Alumina Aluminum oxide (Al_2O_3). Alumina substrates are made of formulations that are primarily alumina.

Ambient temperature Temperature of atmosphere in intimate contact with the electrical part or device.

Analog circuits Circuits that provide a continuous (vs. discontinuous) relationship between the input and output.

Angle of attack The angle between the squeegee face of a thick-film printer and the plane of the screen.

Angled bond Bond impression of first and second bond are not in a straight line.

Angstroms A unit of measurement used in thin film circuits equal to 10^{-10} m.

Annealing Heating of a film resistor followed by slow cooling to relieve stresses and stabilize the resistor material.

Anodization An electrochemical oxidation process used to change the value of thin-film resistors or prepare capacitor dielectrics.

Array A group of elements or circuits arranged in rows and columns on one substrate.

Artwork The accurately scaled configuration or pattern produced usually at an enlarged ratio, to enable the product to be made therefrom by photographic reduction to a 1:1 working pattern; layouts and photographic films which are created to produce the working thick-film screens and thin-film masks.

As-fired Values of thick-film resistors or smoothness of ceramic substrates as they come out of the firing furnace and, respectively, prior to trimming and polishing (if required).

Aspect ratio The ratio between the length of a film resistor and its width; equal to the number of squares of the resistor.

Assembly A film circuit to which discrete components have been attached. Also an assembly of one or more film circuits which may include several components.

Assembly drawing A drawing showing all the components and interconnections mounted or soldered to the film circuit in their proper position. It might also show the assembly of one or more film circuits which may include several discrete components.

Attack angle See Angle of attack.

Axial leads Leads coming out of the ends of a discrete component or device along the central axis rather than out the sides.

Back bonding Bonding active chips to the substrate using the back of the chip, leaving the face, with its circuitry face up. The opposite of back bonding is face down bonding.

Back mounting See Back bonding.

Back radius The radius of the trailing edge of a bonding tool foot.

Backfill Filling an evacuated hybrid circuit package with a dry inert gas prior to hermetically sealing.

Bake out Subjecting an unsealed hybrid circuit package to an elevated temperature to bake out moisture and unwanted gases prior to final sealing.

Ball bond A bond formed when a ball shaped end interconnecting wire is deformed by thermocompression against a metallized pad. The bond is also designated a nail head bond from the appearance of the flattened ball.

Barium titanate ($BaTiO_3$) The basic raw material used to make high-dielectric-constant ceramic capacitors. Used also in high-K thick-film ceramic pastes.

Batch processing Manufacturing method whereby a particular process sequence operates on a large number of components simultaneously.

Bathtub package A boxlike package wherein the substrate is mounted.

Beam lead A long structural member not supported everywhere along its length and subject to the forces of flexure, one end of which is permanently attached to a chip device and the other end intended to be bonded to another material, providing an electrical interconnection or mechanical support or both.

Beam lead device An active or passive chip component possessing beam leads as its primary interconnection and mechanical attachment means to a substrate.

Beryllia Beryllium oxide (BeO). A substrate material used where extremely high thermal conductivity is desired.

Binders Materials added to thick-film compositions and unfired substrates to give sufficient strength for prefire handling.

Bleeding The laterial spreading or diffusion of a printed film into adjacent areas, beyond the geometric dimensions of the printing screen. This may occur during drying, or firing.

Blending Different viscosities of the same types of materials may be blended together to achieve intermediate viscosities. This term is also applied to resistive inks that can be blended with each other to achieve intermediate resistivities.

Blisters Raised parts of a conductor or resistor formed by the outgassing of the binder or vehicle during the firing cycle

Block To plug up open mesh in a screen to prevent resistor and conductor pastes from being deposited in unwanted areas.

Block diagram A circuit diagram in which the essential units of the functional system are drawn in the form of blocks and the relationship between blocks is indicated by appropriate connecting lines.

Block off See Block

Boat A container for materials to be evaporated or fired.

Bond An interconnection which performs a permanent electrical and/or mechanical function.

Bond deformation The change in the form of the lead produced by the bonding tool, causing plastic flow, in making the bond.

Bond envelope The range of bonding parameters over which acceptable bonds may be formed.
Bond interface The interface between the lead and the material to which it was bonded on the substrate.
Bond lift-off The failure mode whereby the bonded lead separates from the surface to which it was bonded.
Bond off See Bond lift-off.
Bond pad See Bonding area.
Bond parameters See Bond schedule.
Bond schedule The values of the bonding machine parameters used when adjusting for bonding. For example, in ultrasonic bonding, the values of the bonding force, time, and ultrasonic power.
Bond site The portion of the bonding areas where the actual bonding took place (see Bonding area).
Bond strength In wire bonding, the pull force at rupture of the bond interface measured in the unit gram-force.
Bond surface See Bonding area.
Bond-to-bond distance The distance measured from the bonding site on the die to the bond impression on the post, substrate land, or fingers, which must be bridged by a bonding wire or ribbon.
Bond-to-chip distance In beam lead bonding the distance from the heel of the bond to the component.
Bond tool The instrument used to position the lead(s) over the desired bonding area and impart sufficient energy to the lead(s) to form a bond.
Bondability Those surface characteristics and conditions of cleanliness of a bonding area which must exist in order to provide a capability for successfully bonding an interconnection material by one of several methods, such as ultrasonic or thermocompression wire bonding.
Bonding, die Attaching the semiconductor chip to the substrate, either with an epoxy, eutectic or solder alloy.
Bonding area The area, defined by the extent of a metallization land or the top surface of the terminal, to which a lead is or is to be bonded.
Bonding island Same as Bonding pad.
Bonding pad A metallized area at the end of a thin metallic strip to which a connection is to be made.
Bonding wire Fine gold or aluminum wire for making electrical connections in hybrid circuits between various bonding pads on the semiconductor device substrate and device terminals or substrate lands.

Borosilicate glass A sealing glass providing a close coefficient of expansion match between some metal leads and ceramic or glass packages.

Braze A joint formed by a brazing alloy (v.). To join metals with a nonferrous filler metal at temperatures above 800°F.

Brazing Similar to soldering. The joining of metals with a nonferrous filler metal at temperatures above 800°F. Also called hard soldering.

Break load See Bond strength.

Breakaway In screen printing the distance between the upper surface of the substrate and the lower surface of the screen when the screen is not deflected by the squeegee.

Breakdown voltage The voltage threshold beyond which there is a marked (almost infinite rate) increase in electrical current conduction.

Bugging height The distance between the hybrid substrate and the lower surface of the beam lead device which occurs because of deformation of beam leads during beam lead bonding.

Bulk conductance Conductance between two points of a homogeneous material.

Burn-in The process of electrically stressing a device (usually at an elevated temperature environment) for an adequate period of time to cause failure of marginal devices.

Burn-off See Flame-off.

Camber A term that describes the amount of overall warpage present in a substrate.

Capacitance density (Also referred to as sheet capacity) A term used to describe the amount of capacitance available per unit area (pF/mil^2 or $\mu F/in.^2$).

Capillary A hollow bonding tool used to guide the bonding wire and to apply pressure to the wire during the bonding cycle.

Capillary tool A tool used in bonding where the wire is fed to the bonding surface of the tool through a bore located along the long axis of the tool.

Catalyst Any substance which affects the rate of chemical reaction, but which itself may be recovered unchanged at the end of the reaction.

Centerline average (CLA) The arithmetical average (AA) of measured deviations in a surface profile from an imaginary mean centerline located between the peaks and valleys. The rms reading for a given surface finish is about 11% higher than the AA reading.

Centerwire break The failure mode in a wire pull test where the wire fractures at approximately midspan.

Centrifuge Testing the integrity of bonds in a hybrid circuit by spinning the circuit at a high rate of speed, thereby imparting a high g loading on the interconnecting wire bonds and bonded elements.

Ceramic Inorganic nonmetallic material such as alumina, beryllia, steatite, or forsterite, whose final characteristics are produced by subjection to high temperatures, often used in microelectronics as parts of components, substrate, or package.
Cermet A solid homogeneous material usually consisting of a finely divided admixture of a metal and ceramic in intimate contact. Cermet thin films are normally combinations of dielectric materials and metals.
Chemical vapor deposition Depositing circuit elements on a substrate by chemical reduction of a vapor on contact with the substrate.
Chessman The disk, knob, or lever used to manually control the position of the bonding tool with respect to the substrate.
Chip The uncased and normally leadless form of an electronic component part, either passive or active, discrete or integrated.
Chip-and-wire A hybrid technology employing face-up-bonded chip devices exclusively, interconnected to the substrate conventionally (i.e., by flying wires).
Chisel A specially shaped bonding tool in the shape of a chisel used for wedge bonding and ultrasonic bonding of aluminum or gold wires to elements or package leads.
Chlorinated hydrocarbon solvents See Halogenated hydrocarbon solvents.
Chopped bond Those bonds with excessive deformation such that the strength of the bond is greatly reduced.
Chuck Portion of the bonding machine that holds the unit to be bonded.
Circuit The interconnection of a number of electrical elements and/or devices, performing a desired electrical function.
Clamping force Force applied to a bonding tool to effect a bond.
Clean room A special manufacturing area where the air is filtered to remove dust particles and precautionary measures are used to keep contamination away from the unprotected circuit during processing.
Clinch A method of mechanically securing components prior to soldering, by bending that portion of the component lead that extends beyond the lip of the mounting hole, against a pad area.
Coefficient The ratio of change under specified conditions of temperature, length, etc.
Coefficient of thermal expansion The ratio of the change in length to the change in temperature.
Co-firing Processing the thick-film conductors and resistors through the firing cycle at the same time.
Coined A screen which contains the impression of a substrate because it has been subject to abuse is said to be coined. The term can also refer to a screen manufactured with a coined impression for the purpose of screening media in a special designed substrate

with a cavity that standard screens cannot achieve. It is also used to describe the process by which the base of a package has been formed.

Coining See Coined.

Cold solder connection A soldered connection where the surfaces being bonded moved relative to one another while the solder was solidifying, causing an uneven solidification structure which may contain microcracks. Such cold joints are usually dull and grainy in appearance.

Cold-weld Forming a hermetic seal in a metal package by welding the lid to the frame using pressure alone.

Collector electrode The metallized bonding pad making ohmic contact with the collector of a transistor element.

Comb pattern A test pattern formed on a substrate in the form of a comb.

Compatible Materials that can be mixed or blended or brought into contact with each other with minimum reaction or separation taking place; or each material added will not degenerate the performance of the whole.

Compensation circuit A circuit which alters the functioning of another circuit to which applied with the goal of achieving a desired performance; temperature and frequency compensation are the most common.

Compensation network Same as Compensation circuit.

Complex arrays An array of integrated devices in which a large number of elements are integral to each device.

Compliant bond A bond which uses an elastically and/or plastically deformable member to impart the required energy to the lead. This member is usually a thin metal foil that is expendable in the process.

Compliant member The elastically and/or plastically deformable medium which is used to impart the required energy to the lead(s) when forming a compliant bond.

Component A diversely used term, which, dependent on context, may mean active or passive element, device, integrated or functional circuit, functional unit, or part of an operating system.

Compound (chemical) A substance consisting of two or more elements chemically united in definite proportions by weight.

Compression seal A seal made between an electronic package and its leads. The seal is formed as the heated metal, when cooled, shrinks around the glass insulator, thereby forming a tight joint.

Conductive adhesive An adhesive material that has metal powder added to increase electrical conductivity.

Conductive epoxy An epoxy material (polymer resin) that has been made conductive by the addition of a metal powder, usually gold or silver.

Conductivity The ability of a material to conduct electricity; the reciprocal of resistivity.

Conductor spacing The distance between adjacent conductor film edges.

Conductor width The width of individual conductors in a conductive film pattern.

Conductors A class of materials that conduct electricity easily, i.e., have low resistivity ($<10^{-4}$ Ω/cm).

Conformal coating A thin nonconductive coating, either plastic (e.g., poly-p-xylylene) or inorganic, applied to a circuit for environmental and/or mechanical protection.

Contact angle The angle made between the bonding material and the bonding pad.

Contact printing A method of screen printing where the screen is almost (within a few mils) in contact with the substrate. Used for printing with metal mask.

Contact resistance In electronic elements, such as capacitors or resistors, the apparent resistance between the terminating electrode and the body of the device.

Continuous belt furnace A firing furnace that has a continuous belt carrying the unfired substrates through the firing cycle.

Controlling collapse Controlling the reduction in height of the solder balls in a flip-chip processing operation.

Coordinatograph A drafting machine of great accuracy used in making original artwork for integrated circuits or microcircuits.

Coplanar leads (flat leads) Ribbon-type leads extending from the sides of the circuit package, all lying in the same plane.

Corona The flow of small erratic current pulses resulting from discharges in voids in a dielectric during voltage stress; also discharge resulting from ionization of gas surrounding a conductor (frequently luminous) which occurs when the potential gradient exceeds a certain value but is not sufficient to cause sparking.

Coupling capacitor A capacitor that is used to block dc signals, and to pass high-frequency signals between parts of an electronic circuit.

Cratering Defect in which portion of chip under ultrasonic bond is torn loose by excessive amount of energy transmitted through the wire bond leaving a pit.

Crazing Fine cracks which may extend on or through layers of plastic of glass materials.

Creep The dimensional change with time of a material under load.

Crossover The transverse crossing of metallization paths without mutual electrical contact and achieved by the deposition of an insulating layer between the conducting paths at the area of crossing.

Crosstalk Signals from one line leaking into another nearby conductor because of capacitive or inductive coupling or both (e.g., owing to the capacitance of a thick-film crossover).

Crystal growth The formation of crystals in a material over a period of time and at an established temperature.

Cure time The total elapsed time between the addition of a catalyst and the complete hardening of a material; also the time for hardening of premixed, frozen, or refrigerated epoxy adhesives.

Curie temperature (Curie point) Above a critical temperature, ferromagnetic materials lose their permanent spontaneous magnetization and ferroelectric materials lose their spontaneous polarization. This critical temperature is the Curie point. At this point, ferroelectric ceramic capacitors reach a peak in capacitance.

Curing agent A material which when added to a second material activates a catalyst already present in the second material, thereby bringing about a chemical reaction, usually causing a hardening of the entire mass.

Curls Extruded material coming out from edge of bond.

Current carrying capacity The maximum current which can be continuously carried by a circuit without causing objectionable degradation of the electrical or mechanical properties.

Custom circuits Circuits designed to satisfy a single application requirement (hybrid or monolithic).

Cut and strip A method of producing artwork using a two-ply laminated plastic sheet, by cutting and stripping off the unwanted portions of the opaque layer from the translucent layer, leaving the desired art work configuration.

Cutoff The operation following the final bonding step that separates the bond from the wire magazine.

Cutoff scissors The scissors on a bonder to sever the wire after bonding.

Cyclic stress A completed circuit subjected to stress by cycling temperature and load over a period of time to cause premature failure.

Dc voltage coefficient The measure of changes in the primary characteristics of a circuit element as a function of the voltage stress applied.

Definition The sharpness of a screen printed pattern—the exactness with which a pattern is printed.

Degradation Change for the worse in the characteristics of an electric element because of heat, high voltage, etc.

Deionized water Water that has been purified by removal of ionizable materials.

Delta (Δ) limit The maximum change in a specified parameter reading which will permit a hybrid microcircuit to be accepted on the specified test, based on a comparison of the fianl measurement with a specified previous measurement. (*Note*: When expressed as a percentage value, it shall be calculated as a proportion of the previous measured value.)

Density Measured as mass per unit volume or the weight of a material in relationship to its volume or size.

Detritus Loose material, dislodged during resistor trimming but remaining in the trimmed area.

Device A single discrete electronic element such as a transistor or resistor, or a number of elements integrated within one die, which cannot be further reduced or divided without eliminating its stated function. Preferred usage is die or dice.

Dewetting The condition in a soldered area in which liquid solder has not adhered intimately and has pulled back from the conductor area.

Dice The plural of die.

Die An uncased discrete or integrated device obtained from a semiconductor wafer (*see* Chip).

Die bond Attachment of a die or chip to the hybrid substrate.

Dielectric Materials that do not conduct electricity and that are used for making capacitors for insulating conductors (as in crossover and multilayered circuits), and for encapsulating circuits.

Dielectric breakdown The breakdown of the insulation resistance in a medium under high voltage.

Dielectric constant The term used to describe a material's ability to store charge when used as a capacitor dielectric. It is the ratio of the charge that would be stored with free space as the dielectric to that stored with the material in question as the dielectric.

Dielectric layer A layer of dielectric material between two conductor plates.

Dielectric loss The power dissipated by a dielectric as the friction of its molecules opposes the molecular motion produced by an alternating electric field.

Dielectric properties The electrical properties of a material such as insulation resistance, breakdown voltage, etc.

Dielectric strength The maximum electric field that a dielectric will withstand without breaking down (physically). Expressed in volts per unit distance, such as centimeter, mil, etc.

Diffusion The phenomenon of movement of matter at the atomic level from regions of high concentration to regions of low concentration.

Diffusion bond *See* Solid-phase bond.

Diffusion constant The relative rate at which diffusion takes place with respect to temperature.

Digital circuits Applied normally for switching applications where the output of the circuit normally assumes one or two states (binary operation); however, three state operation is possible.

Direct contact A contact made to the semiconductor die when the wire is bonded directly over the part to be electrically connected, as opposed to the expanded contact.

Direct emulsion Emulsion applied to a screen in a liquid form as contrasted to an emulsion that is transferred from a backing film of plastic.

Direct emulsion screen A screen whose emulsion is applied by painting directly onto the screen, as opposed to indirect emulsion type.

Direct metal mask A metal mask made by etching a pattern into a sheet of metal.

Discrete As applied to components used in thin- and thick-film hybrid circuits; the elements that are added separately are discrete elements (or devices) as opposed to those that are made by screen printing or vacuum deposition methods as parts of the film network.

Discrete components Individual components such as resistors, capacitors, and transistors.

Dissipation factor Tangent of the dielectric loss angle. Dissipation factor is the ratio of the resistive component of a capacitor (R_c) to the capacitive reactance (X_c) of the capacitor.

Doping The addition of an impurity to a semiconductor to alter its conductivity.

Drift Permanent change in value of a device parameter over a period of time because of the effects of temperature, aging, humidity, etc.

Dry air Air that has been circulated through a drying process to remove water molecules.

Dry inert atmosphere An inert gas such as nitrogen that has been circulated through a drying process to remove water molecules.

Dry pressing Pressing and compacting of dry powdered materials with additives together in rigid die molds under heat and pressure to form a solid mass, usually followed by sintering as for alumina substrates.

Dry print The screened resistor and conductors that have gone through the drying cycle removing the solvents from the ink.

Dryer A drying tube containing silica gel or a similar moisture absorbent chemical.

Dual-in-line pack (DIP) A package having two rows of leads extending at right angles from the base and having standard spacings between leads and between rows of leads.

Ductility That property which permits a material to deform plastically without fracture.

Dynamic printing force The fluid force which causes a pseudoplastic paste to flow through a screen mesh and wet the surface beneath. Its absolute value is a complex function of all screen printer operating parameters together with the rheological properties of the fluid being printed.

Dynamic testing Testing a hybrid circuit where reactions to ac (especially high frequency) are evaluated.
Edge definition See Definition.
Electric field A region where there is a voltage potential, the potential level changing with distance. The strength of the field is expressed in volts per unit distance.
Electrical isolation Two conductors isolated from each other electrically by an insulating layer
Electrical properties The properties of a device or material that effect its conductivity or resistivity to the flow of an electric current.
Electrically hot case A hybrid circuit package that is used as part of the grounding circuit.
Electrodes The conductor or conductor lands of a hybrid circuit. Also the metallic portions of a capacitor structure.
Electroless plating Deposit of a metallic material on a surface by chemical deposition as opposed to the use of an electric current.
Electron beam bonding Bonding two conductors by means of heating with a stream of electrons in a vacuum.
Electronic packaging The technical discipline of designing a protective enclosure for an electronic circuit so that it will both survive and perform under a plurality of environmental conditions.
Element A constituent unit which contributes to the operation of a hybrid microcircuit. Integral elements include deposited or screened passive circuit elements, metallization paths, and deposited or formed insulation. Discrete elements include discrete or integrated electronic parts, chips, and interconnecting wires or ribbon.
Elongation The ratio of the increase in wire length at rupture, in a tensile test, to the initial length, given in percent.
Embedded Enclosed in a plastic material.
Emitter electrode The metallic pad making ohmic contact to the emitter area of a transistor element.
Emulsion The light-sensitive material used to coat the mesh of a screen.
Encapsulate Sealing up or covering an element or circuit for mechanical and environmental protection.
Entrapped material Gas or particles bound up in an electrical package that cannot escape.
Environmental test A test or series of tests used to determine the sum of external influences affecting the structural, mechanical, and functional integrity of any given package or assembly.
Etched metal mask A metal mask used for screening wherein the pattern is created in a sheet of metal by the etching process.
Eutectic (1) A term applied to the mixture of two or more substances that has the lowest melting point. (2) An alloy or solution having its components in such proportion that the melting point is the lowest possible with those components.

Glossary of Terms 177

Eutectic alloy An alloy having the same temperature for melting and solidus.
Expanded contact A contact made to the semiconductor die where the wire bonded to an area remote from the part to be electrically connected so that a lateral interconnection path for the current is required.
Exponential failures (wear-out) Failures that occur at an exponentially increasing rate.
External leads Electronic package conductors for input and output signals, power, and ground. Leads can be either flat ribbons or round wires.
Eyelet tool A bonding tool with a square protuberance beneath the bonding tool surface which presses into the conductor and prevents the slippage between wire or conductor and tool interface. Used primarily for ribbon wire bonding.
Face bonding The opposite of back bonding. A face bonded semiconductor chip is one that has its circuitry side facing the substrate. Flip-chip and beam lead bonding are the two common face bonding methods.
Failure analysis The analysis of a circuit to locate the reason for the failure of the circuit to perform to the specified level.
Failure mechanism The physical or chemical process by which a device proceeds to the point of failure.
Failure mode The cause for rejection of any failed device as defined in terms of the specific electrical or physical requirement that it failed to meet.
Failure rate The rate at which devices from a given population can be expected (or were found) to fail as a function of time (e.g., %/1000 hr of operation).
Fatigue Used to describe a failure of any structure caused by repeated application of stress over a period of time.
Fatigue factor The factor causing the failure of a device under repeated stress.
Feathers See Curls.
Feedthrough A conductor through the thickness of a substrate, thereby electrically connecting both surfaces.
Ferrite A powdered, compressed, and sintered magnetic material having high resistivity; cores made of sintered powders are used for ferromagnetic applications.
Ferroelectric A crystalline dielectric that exhibits dielectric hysteresis—an electrostatic analogy to ferromagnetic materials.
Ferromagnetic A material that has a relative permeability noticeably exceeding unity and generally exhibits hysteresis.
Field trimming Trimming of a resistor to set an output voltage, current, etc.

Filled plastic Adding ceramic, silica, or metal powder to an encapsulant to improve the thermal conductivity.

Filler A substance, usually dry and powdery or granular, used to thicken fluids or polymers.

Fillet A concave junction formed where two surfaces meet.

Film Single or multiple layers or coatings of thin- or thick-material used to form various elements (resistors, capacitors, inductors) or interconnections and crossovers (conductors, insulators). Thin films are deposited by vacuum evaporation or sputtering and/or plating. Thick films are deposited by screen printing.

Film conductor A conductor formed in situ on a substrate by depositing a conductive material by screening, plating, or evaporation techniques.

Film integrated circuit See Film microcircuit.

Film microcircuit The final form of an electrical circuit function utilizing a thick- or a thin-film network.

Film network An electrical network composed of thin- and/or thick-film components and interconnections deposited on a substrate.

Final seal The manufacturing operation that completes the enclosure of the hybrid microcircuit so that further internal processing cannot be performed without debidding or disassembling the package.

Fine leak A leak in a sealed package less than $10^{-5} cm^3/sec$ at one atmosphere of differential air pressure.

Fire The term used to describe the act of heating a thick-film circuit so that the resistors, conductors, capacitors, etc., will be transformed into their final form.

Firing sensitivity Refers to the percentage change caused in the fired film characteristics due to a change in peak firing temperature. The firing sensitivity is expressed in units of $\%/°C$.

First bond The first bond in a sequence of two or more bonds made to form a conductive connection.

First radius The radius of the front edge of a bonding tool foot.

First search That period of machine cycle at which final adjustment in the location of the first bonding area (see first bond) under the tool are made prior to lowering the tool to make the first bond.

Fissuring The cracking of dielectric or conductors. Often dielectrics, if incorrectly processed, will crack in the presence of conductors because of stresses occurring during firing.

Flag Support area on lead frame for die.

Flame-off The procedure where the wire is severed by passing a flame across the wire, thereby melting it. The procedure is used in gold wire thermocompression bonding to form a ball for making a ball bond.

Flat pack An integrated circuit package having its leads extending from the sides and parallel to the base.

Flexible coating A plastic coating that is still flexible after curing.
Flexural strength Strength of the laminate when it is measured by bending.
Flip chip A leadless, monolithic structure, containing circuit elements, which is designed to electrically and mechanically interconnect to the hybrid circuit by means of an appropriate number of bumps located on its face which are covered with a conductive bonding agent.
Flip-chip mounting A method of mounting flip chips on thick- or thin-film circuits without the need for subsequent wire bonding.
Floating ground An electrical ground circuit that does not allow connection between the power and signal ground for the same circuit.
Flood bar A bar or other such device on a screen printing device that will drag paste back to the starting point after the squeegee has made a printing stroke. The flood stroke returns the paste without pushing it through the meshes, so it does no printing, only returns the paste supply to be ready for the next print.
Flux In soldering, a material that chemically attacks surface oxides and tarnishes so that molten solder can wet the surface to be soldered.
Flux residue Particles of flux remaining on a circuit after soldering and cleaning operations.
Foot length The long dimension of the bonding surface of a wedge-type bonding tool.
Footprint The area needed on a substrate for a component or element. Usually refers to specific geometric pattern of a chip.
Forming gas A gas (N_2, with traces of H_2 and He) used to blanket a part being processed to prevent oxidation of the metal areas.
Frit Glass composition ground up into a powder form and used in thick-film compositions as the portion of the composition that melts upon firing to give adhesion to the substrate and hold the composition together.
Functional trimming Trimming of a circuit element (usually resistors) on an operating circuit to set a voltage or current on the output.
Furnace active zone The thermostatically controlled portion of a multizoned muffle furnace.
Furnace profile See Firing profile.
Furnace slave zone That portion of a multizoned muffle furnace where the instantaneous power supplied to the heating element is a set percentage of the power supplied to the active zone. Hence temperature control in the slave zone is not accomplished by sensing thermocouples as in the case of the active zone.
Fusing Melting and cooling two or more powder materials together so that they bond together in a homogeneous mass.

Gas blanket An atmosphere of inert gas, nitrogen, or forming gas flowing over a heated integrated circuit chip or a substrate that keeps the metallization from oxidizing during bonding.

Glass binder The glass powder added to a resistor or conductor ink to bind the metallic particles together after firing.

Glass phase The part of the firing cycle wherein the glass binder is in a molten phase.

Glassivation A method of semiconductor passivation by coating the element with a pyrolytic glass deposition.

Glaze See Overglaze.

Glazed substrate A glass coating on a ceramic substrate to effect a smooth and nonporous surface.

Glossy A shiny surface usually formed by the glass matrix in a conductor or resistor ink.

Grain growth The increase in the size of the crystal grains in a glass coating or other material over a period of time.

Green A term used in ceramic technology meaning unfired. For example, a "green" substrate is one that has been formed, but has not been fired.

Gross leak A leak in a sealed package greater than $10^{-5} cm^3/sec$ at one atmosphere of differential air pressure.

Ground plane A conductive layer on a substrate or buried within a substrate that connects a number of points to one or more grounding electrodes.

Halo effect A glass halo around certain conductors. Generally, this is an undesirable effect to be avoided by changing furnace profiles or material types.

Halogenated hydrocarbon solvents Organic solvents containing the elements chlorine or fluorine used in cleaning substrates and completed circuits (e.g., trichloroethylene, various Freons, etc.).

Hand soldered Forming a soldered connection with solder using a hand-held soldering iron for application of the heat.

Hard solder Solder that has a melting point above 800 °F (425°C).

Hardness A property of solids, plastics, and viscous liquids that is indicated by their solidity and firmness; resistance of a material to indentation by an indentor of fixed shape and size under a statis load or to scratching; ability of a metal to cause rebound of a small standard object dropped from a fixed height; the cohesion of the particles on the surface of a mineral as determined by its capacity to scratch another or be itself scratched.

Header The base of a hybrid circuit package that holds the leads.

Heat clean The process of removing all organic material from glass cloth to approximately 650 to 700°F for a period of time ranging up to 50 hr.

Heat column The heating element in a eutectic die bonder or wire bonder used to bring the substrate up to the bonding temperature.

Heat flux The outward flow of heat from a heat source.
Heat sink The supporting member to which electronic components or their substrate or their package bottom are attached. This is usually a heat conductive metal with the ability to rapidly transmit heat from the generating source (component).
Heat soak Heating a circuit over a period of time to allow all parts of the package and circuit to stabilize at the same temperature.
Heel (of the bond) The part of the lead adjacent to the bond that has been deformed by the edge of the bonding tool used in making the bond. The back edge of the bond.
Heel break The rupture of the lead at the heel of the bond.
Heel crack A crack across the width of the bond in the heel region.
Hermetic Sealed up so that it is gastight. The test for hermeticity is to observe leak rates when placed in a vacuum. A plastic encapsulation cannot be hermetic by definition because there is no internal volume of gas to escape.
Hermeticity The ability of a package to prevent exchange of its internal gas with the external atmosphere. The figure of merit is the gaseous leak rate of the package measured in $atm\text{-}cm^3/sec$.
Hi-K Abbreviation for high dielectric constant.
High-K ceramic A ceramic dielectric composition (usually $BaTiO_3$) which exhibits large dielectric constants, and nonlinear voltage and temperature response.
High-purity alumina Alumina having over 99% purity of Al_2O_3.
Hole diameter Normally refers to the diameter of the hole through the bonding tool.
Homogeneous Alike or uniform in composition. A thick-film composition that has settled out is not homogeneous, but after proper stirring it is. The opposite of heterogeneous.
Horn Cone-shaped member which transmits ultrasonic energy from transducer to bonding tool.
Hostile environment An environment that has a degrading effect on an electronic circuit.
Hot spot A small area on a circuit that is unable to dissipate the generated heat and therefore operates at an elevated temperature above the surrounding area.
Hot zone The part of a continuous furnace or kiln that is held at maximum temperature. Other zones are the preheat zone and cooling zone.
Hybrid circuit A microcircuit consisting of elements which are a combination of the film circuit type and the semiconductor circuit type, or a combination of one or both of these types and may include discrete add-on components.
Hybrid group (electrically and structurally similar circuits) Hybrid microcircuits which are designed to perform the same type(s) of basic circuit function(s), for the same supply, bias, and single

voltages and for input-output compatibility with each other under an established set of loading rules, and which are enclosed in packages of the same construction and outline.

Hybrid integrated circuit A microcircuit including thick-film or thin-film paths and circuit elements on a supporting substrate, to which active and passive microdevices are attached, either prepackaged or in uncased form as chips, usually all enclosed in a suitable package (hermetic or epoxy type). Used interchangeably with hybrid circuit and hybrid microcircuit.

Hybrid microcircuit A microcircuit that involves an insulating substrate on which are deposited networks, consisting generally of conductors, resistors, and capacitors, and to which are attached discrete semiconductor devices and/or monolithic integrated circuits and/or passive elements to form a packaged assembly. Used interchangeably with hybrid circuit and hybrid integrated circuit.

Hybrid microelectronics The entire body of electronic art which is connected with or applied to the realization of electronic systems using hybrid circuit technology.

Hybrid microwave circuit See Microwave integrated circuit.

Imbedded layer A conductor layer having been deposited between insulating layers.

Inactive flux Flux that becomes nonconductive after being subjected to the soldering temperature.

Inclined plane furnace A resistor firing furnace having the hearth inclined so that a draft of oxidizing atmosphere will flow through the heated zones through natural convection means.

Incomplete bond A bond impression having dimensions less than normal size due to a portion of the bond impression being missing.

Indirect emulsion Screen emulsion that is transferred to the screen surface from a plastic carrier or backing material.

Indirect emulsion screen A screen whose emulsion is a separate sheet or film of material, attached by pressing into the mesh of the screen (as opposed to the direct emulsion type).

Inert atmosphere A gas atmosphere such as helium or nitrogen that is nonoxidizing or nonreducing of metals.

Infant mortality (early failures) The time regime during which hundreds of circuits are failing at a decreasing rate (usually during the first few hundred hours of operation).

Infrared The band of electromagnetic wavelengths lying between the extreme of the visible (≈ 0.75 μm) and the shortest microwaves (≈ 1000 μm). Warm bodies emit the radiation and bodies which absorb the radiation are warmed.

Injection molded Molding of electronic packages by injecting liquefied plastic into a mold.

Ink Synonymous with "composition" and "paste" when relating to screenable thick-film materials, usually consisting of glass frit, metals, metal oxide, and solvents.

Ink blending See Blending.

In-process Some step in the manufacturing operation prior to final testing.

Insertion loss The difference between the power received at the load before and after the insertion of apparatus at some point in the line.

Inspection lot A quantity of hybrid microcircuits, representing a production lot, submitted for inspection at one time to determine compliance with the requirements and acceptance criteria of the applicable procurement document. Each inspection lot consists of one or more inspection sublots of electrically and structurally similar circuits.

Inspection sublot A quantity of hybrid microcircuits, which is part of or an entire inspection lot, submitted for inspection at one time to determine compliance with the requirements and acceptance criteria of the applicable procurement document. Each inspection sublot of circuits should be a group identified as having common manufacturing experience through all significant manufacturing operations.

Insulation resistance (IR) The resistance to current flow when a potential is applied. IR is measured in megohms.

Insulators A class of materials with high resistivity. Materials that do not conduct electricity. Materials with resistivity values of over 10^5 Ω/cm are generally classified as insulators.

Integrated circuit A microcircuit (monolithic) consisting of interconnected elements inseparably associated and formed in situ on or within a single substrate (usually silicon) to perform an electronic circuit function.

Interconnection The conductive path required to achieve connection from a circuit element to the rest of the circuit.

Interface The boundary between dissimilar materials, such as between a film and substrate or between two films.

Interfacial bond An electrical connection between the conductors on the two faces of a substrate.

Intermetallic bond The ohmic contact made when two metal conductors are welded or fused together.

Intermetallic compound A compound of two or more metals that has a characteristic crystal structure that may have a definite composition corresponding to a solid solution, often refractory.

Internal visual See Preseal visual.

Intraconnections Those connections of conductors made within a circuit on the same substrate.

Ion migration The movement of free ions within a material or across the boundry between two materials under the influence of an applied electric field.

Ionizable material Material that has electrons easily detracted from atoms or molecules, thus originating ions and free electrons that will reduce the electrical resistance of the material.

Jumper A direct electrical connection between two points on a film circuit. Jumpers are usually portions of bare or insulated wire mounted on the component side of the substrate.

Junction temperature The temperature of the region of transition between the p- and n-type semiconductor material in a transistor or diode element.

K Symbol for dielectric constant.

K factor This term refers to thermal conductivity, the ability of a substance to conduct heat through its mass.

Kerf The slit of channel cut in a resistor during trimming by laser beam or abrasive jet.

Kiln A high-temperature furnace used in firing ceramics.

Kirkendall voids The formation of voids by diffusion across the interface between two different materials, in the material having the greater diffusion rate into the other.

L cut A trim notch in a film resistor that is created by the cut starting perpendicular to the resistor length and turning 90° to complete the trim parallel to the resistor axis thereby creating an L-shaped cut.

Ladder network A series of film resistors with values from the highest to the lowest resistor reduced in known ratios.

Laminar flow A constant and directional flow of filtered air across a clean workbench. The flow is usually parallel to the surface of the bench.

Laminate A layered sandwich of sheets of substances bonded together under heat and pressure to form a single structure.

Lands Widened conductor areas on the major substrate used as attachment points for wire bonds or the bonding of chip devices.

Lapping Smoothing a substrate surface by moving it over a flat plate having a liquid abrasive.

Laser bonding Effecting a metal-to-metal bond of two conductors by welding the two materials together using a laser beam for a heat source.

Laser trim The adjustment (upward) of a film resistor value by applying heat from a focused laser source to remove material.

Lattice structure A stable arrangement of atoms and their electron-pair bonds in a crystal.

Layer One of several films in a multiple film structure on a substrate.

Layout The positioning of the conductors and/or resistors on artwork prior to photoreduction of the layout to obtain a working negative or positive used in screen preparation.
Leaching In soldering, the dissolving (alloying) of the material to be soldered into the molten solder.
Lead A conductive path which is usually self-supporting.
Lead frames The metallic portion of the device package that completes the electrical connection path from the die or dice and from ancillary hybrid circuit elements to the outside world.
Lead wires Wire conductors used for intraconnections or input/output leads.
Leadless device A chip device having no input/output leads.
Leadless inverted device See LID.
Leakage current An undesirable small stray current which flows through or across an insulator between two or more electrodes, or across a back-biased junction.
Leveling A term describing the settling or smoothing out of the screen mesh marks in thick films that takes place after a pattern is screen printed.
LID Abbreviation for "leadless inverted device." A shaped, metallized ceramic form used as an intermediate carrier for semiconductor chip devices, especially adapted for attachment to conductor lands of a thick- or thin-film network by reflow solder bonding.
Life drift The change in either absolute level or slope of a circuit element under load. Rated as a percentage change from the original value per 1000 hr of life.
Life test Test of a component or circuit under load over the rated life of the device.
Lift-off mark Impression in bond area left after lift-off removal of a bond.
Line certification Certification that a production line process sequence is under control and will produce reliable circuits in compliance with requirements of applicable mandatory documents.
Line definition A descriptive term indicating a capability of producing sharp, clean screen printed lines. The precision of line width is determined by twice the line edge definition/line width. A typical precision of 4% exists when the line edge definition/line width is 2%.
Linear circuits A circuit with an output that changes in magnitude with relation to the input as defined by a constant factor (*See* Analog circuits).
Lines Conductor runs of a film network.
Liquidus The line on a phase diagram above which the system has molten components. The temperature at which melting starts.

Load life The extended period of time over which a device can withstand its full power rating.
Loop The curve or arc by the wire between the attachment points at each end of a wire bond.
Loop height A measure of the deviation of the wire loop from the straight line between the attachment points of a wire bond. Usually, it is the maximum perpendicular distance from this line to the wire loop.
Loss tangents The decimal ratio of the irrecoverable to the recoverable part of the electrical energy introduced into an insulating material by the establishment of an electric field in the material.
Low-loss substrate A substrate with high radio-frequency resistance and hence slight absorption of energy when used in a microwave integrated circuit.
Mask The photographic negative that serves as the master for making thick-film screens and thin-film patterns.
Mass spectrometer An instrument used to determine the leak rate of a hermetically sealed package by ionizing the gas outflow permitting an analysis of the flow rate in cm^3/sec at 1 atm differential pressure.
Master batch principle Blending resistor pastes to a nominal value of Ω/square. The nominal value is the master control number.
Master layout The original layout of a circuit.
Matte finish A surface finish on a material that has a grain structure and diffuses reflected light.
Mesh size The number of openings per inch in a screen. A 200-mesh screen has 200 openings per linear inch, 40,000 openings per square inch.
Metal inclusion Metal particles embedded in a nonmetal material such as a ceramic substrate.
Metal mask (screens) A screen made not from wire or nylon thread but from a solid sheet of metal in which holes have been etched in the desired circuit pattern. Useful for precision and/or fine printing and for solder cream printing.
Metallization A film pattern (single or multilayer) of conductive material deposited on a substrate to interconnect electronic components, or the metal film on the bonding area of a substrate which becomes a part of the bond and performs both an electrical and a mechanical function.
Metal-to-glass seal (or glass-to-metal seal) An insulating seal made between a package lead and the metal package by forming a glass bond to oxide layers on both metal parts. In this seal, the glass has a coefficient of expansion that closely matches the metal parts.
Microbond A bond of a small wire such as 0.001-in.-diameter gold to a conductor or to a chip device.

Microcircuit A small circuit (hybrid or monolithic) having a relatively high equivalent circuit element density, which is considered as a single part on (hybrid) or within (monolithic) a single substrate to perform an electronic circuit function. (This excludes printed wiring boards, circuit card assemblies, and modules composed exclusively of discrete electronic parts.)

Microcircuit module An assembly of microcircuits or an assembly of microcircuits and discrete parts, designed to perform one or more electronic circuit functions, and constructed such that for the purposes of specification testing, commerce, and maintenance, it is considered indivisible.

Microcomponents Small discrete components such as chip transistors and capacitors.

Microcracks A thin crack in a substrate or chip device, or in thick-film trim-kerf walls, that can only be seen under magnification and which can contribute to latent failure phenomena.

Microelectronics That area of electronic technology associated with or applied to the realization of electronic systems from extremely small electronic parts or elements.

Micron An obsolete unit of length equal to a micrometer (μm).

Micropositioner An instrument used in positioning a film substrate or device for bonding or trimming.

Microprobe A small sharp-pointed probe with a positioning handle used in making temporary ohmic contact to a chip device or circuit for testing.

Microstrip A microwave transmission component usually on a ceramic substrate.

Microstructure A structure composed of finely divided particles bound together.

Microwave integrated circuit A miniature microwave circuit usually using hybrid circuit technology to form the conductors and attach the chip devices.

Migration An undesirable phenomenon whereby metal ions, notably silver, are transmitted through another metal, or across an insulated surface, in the presence of moisture and an electrical potential.

Mil A unit equal to 0.001 in. or 0.0254 mm.

Mislocated bond See Off bond.

Moisture stability The stability of a circuit under high-humidity conditions such that it will not malfunction.

Monocrystalline structure The granular structure of crystals which have uniform shapes and arrangements.

Monolithic ceramic capacitor A term sometimes used to indicate a multi-layer ceramic capacitor.

Monolithic integrated circuit An integrated circuit consisting of elements formed in situ on or within a semiconductor substrate with at least one of the elements formed within the substrate.

MOS device Abbreviation for a metal oxide semiconductor device.

Mother board A circuit board used to interconnect smaller circuit boards called "daughter boards."

MTBF Abbreviation for mean time between failures. A term used to express the reliability level. The reciprocal of the failure rate.

Multichip integrated circuit An integrated circuit whose elements are formed on or within two or more semiconductor chips which are separately attached to a substrate or header.

Multichip microcircuit A microcircuit consisting solely of active dice and passive chips which are separately attached to the major substrate and interconnected to form the circuit.

Multilayer ceramic capacitor A miniature ceramic capacitor manufactured by paralleling several thin layers of ceramic. The assembly is fired after the individual layers have been electroded and assembled.

Multilayer circuits A composite circuit consisting of alternate layers of conductive circuitry and insulating materials (ceramic or dielectric compositions) bonded together with the conductive layers interconnected as required.

Multilayer substrates Substrates that have buried conductors so that complex circuitry can be handled. Assembled using processes similar to those used in multilayer ceramic capacitors.

Multiple circuit layout Layout of an array of identical circuits on a substrate.

Nailhead bond See Ball bond.

Neck break A bond breaking immediately above gold ball of a thermocompression bond.

Negative image The reverse print of a circuit.

Negative temperature coefficient The device changes its value in the negative direction with increased temperature.

Noble metal paste Paste materials composed partially of noble metals such as gold or ruthenium.

Noise Random small variations in voltage or current in a circuit due to the quantum nature of electronic current flow, thermal considerations, etc.

Noise characteristic The characteristics of a resistor to emit unwanted signals within a dynamic electrical system.

Nominal resistance value The specified resistance value of the resistor at its rated load.

Nonconductive epoxy An epoxy material (polymer resin) either without a filler or with a ceramic powder filler added for increasing thermal conductivity and improving thixotropic properties. Nonconductive epoxy adhesives are used in chip to substrate bonds where electri-

Glossary of Terms

cal conductivity to the bottom of the chip is unnecessary or in substrate-to-package bonding.

Nonlinear dielectric A capacitor material that has a nonlinear capacitance voltage relationship. Titanate (usually barium titanate) ceramic capacitors (Class II) are nonlinear dielectrics. NPO and Class I capacitors are linear by definition.

NPO A commonly used code that is synonymous with EIA code COG. Both indicate a temperature characteristic of ± 30 ppm/°C or less for a capacitor.

Nugget The region of recrystallized material at a bond interface which usually accompanies the melting of the materials at the interface.

Occluded contaminants Contaminants that have been adsorbed by a material.

Off bond Bond that has some portion of the bond area extending off the bonding pad.

Off contact (screen printing) The opposite of contact printing in that the printer is set up with a space between the screen and the substrate and contacts the substrate only when the squeegee is cycled across the screen.

Off-contact screener A screener machine that uses off contact printing of patterns onto substrates.

Ohmic contact A contact that has linear voltage current characteristics throughout its entire operating range.

Ohms/square The unit of sheet resistance, or more properly, of sheet resistivity.

Operator certification A program wherein an operator has been qualified and certified to operate a machine.

Organic flux A flux composed of rosin base and a solvent.

Organic vehicle The organic vehicle in a flux is the rosin base material.

Outgas The release of gas from a material over a period of time.

Overbonding See Chopped bond.

Overcoat A thin film of insulating material, either plastic or inorganic (e.g., glass or silicon nitride) applied over integral circuit elements for the purposes of mechanical protection and prevention of contamination.

Overglaze A glass coating that is over another component or element, normally for physical or electrical protection purposes.

Overlap The contact area between a film resistor and a film conductor.

Overlay One material applied over another material.

Overspray The unwanted spreading of the abrasive material coming from the trim nozzle in a resistor trimming machine. The overspray affects the values of adjacent resistors not intended to be trimmed.

Overtravel Overtravel is the excess downward distance a squeegee blade would push the screen if the substrate were not in position.

Oxidizing atmosphere An air-, or other oxygen-containing atmosphere in a firing furnace which oxidizes the resistor materials while they are in the molten state, thereby increasing their resistance.

Package The container for an electronic component(s) with terminals to provide electrical access to the inside of the container. In addition, the container usually provides hermetic and environmental protection for, and a particular form factor to, the assembly of electronic components.

Package cap The cuplike cover that encloses the package in the final sealing operation.

Package lid A flat cover plate that is used to seal a package cavity.

Pad A metallized area on the surface of an active substrate as an integral portion of the conductive interconnection pattern to which bonds or test probes may be applied (*see* Lands for passive substrate).

Parallel gap solder Passing a high current through a high-resistance gap between two electrodes to remelt solder thereby forming an electrical connection.

Parallel gap weld Passing a high current through a high-resistance gap between two electrodes that are applying force to two conductors, thereby heating the two workpieces to the welding temperature and effecting a welded connection.

Parallelism (substrate) The degree of variation in the uniform thickness of a given substrate.

Parasite losses Losses in a circuit often caused by the unintentional creation of capacitor elements in a film circuit by conductor crossovers.

Partial lift A bonded lead partially removed from the bonded area.

Passivated region Any region covered by glass, SiO_2, nitride, or other protective material.

Passivation The formation of an insulating layer directly over a circuit or circuit element to protect the surface from contaminants, moisture, or particles.

Passive components (elements) Elements (or components) such as resistors, capacitors, and inductors which do not change their basic character when an electrical signal is applied. Transistors and electron tubes are active components.

Passive network A circuit network of passive elements such as film resistors that are interconnected by conductors.

Passive substrate A substrate that serves as a physical support and thermal sink for a film circuit that does not exhibit transistance.

Paste Synonymous with "composition" and "ink" when relating to screenable thick-film materials.

Paste blending Mixing a resistor pastes of different Ω/square value to create a third value in between those of the two original materials.

Paste soldering Finely divided particles of solder suspended in a flux paste. Used for screening application onto a film circuit and reflowed to form connections to chip components.

Paste transfer The movement of a resistor, conductor, or solder paste material through a mask and deposition in a pattern onto a substrate.

Pattern The outline of a collection of circuit conductors and resistors that defines the area to be covered by the material on a film circuit substrate.

Peak firing temperature The maximum temperature seen by the resistor or conductor paste in the firing cycle as defined by the firing profile.

Peel bond Similar to lift-off of the bond with the idea that the separation of the lead from the bonding surface proceeds along the interface of the metallization and substrate insulation rather than the bond-metal interface.

Peel strength (peel test) A measure of adhesion between a conductor and the substrate. The test is performed by pulling or peeling the conductor off the substrate and observing the force required. Units are oz/mil or lb/in. of conductor width.

Percent defective allowable (PDA) The maximum observed percent defective which will permit the lot to be accepted after the specified 100% test.

Perimeter sealing area The sealing surface on an electronic package that follows the perimeter of the package cavity and defines the area used in bonding to the lid or cap.

Phase (As glassy phase, or metal phase) Refers to the part or portion of materials system that is metallic or glassy in nature. A phase is a structurally homogeneous physically distinct portion of a substance or a group of substances which are in equilibrium with each other.

Phase diagram State of a metal alloy over a wide temperature range. The phase diagram is used to identify eutectic solders and their solidus/liquidus point.

Photo etch The process of forming a circuit pattern in metal film by light hardening a photosensitive plastic material through a photo negative of the circuit and etching away the unprotected metal.

Photoresist A photosensitive plastic coating material which when exposed to UV light becomes hardened and is resistant to etching solutions. Typical use is as a mask in photochemical etching of thin films.

Pigtail A term that describes the amount of excess wire that remains at a bond site beyond the bond. Excess pigtail refers to remnant wire in excess of three wire diameters.

Pinhole Small holes occurring as imperfections which penetrate entirely through film elements, such as metallization films or dielectric films.

Pits Depressions produced in metal or ceramic surfaces by nonuniform deposition.

Plastic A polymeric material, either organic (e.g., epoxy) or silicone used for conformal coating, encapsulation, or overcoating.

Plastic device A device wherein the package, or the encapsulant material for the semiconductor die, is plastic. Such materials as epoxies, phenolics, silicones, etc., are included.

Plastic encapsulation Environmental protection of a completed circuit by embedding it in a plastic such as epoxy or silicone.

Plastic shell A thin plastic cup or box used to enclose an electronic circuit for environmental protection or used as a means to confine the plastic encapsulant used to imbed the circuit.

Plug-in-package An electronic package with leads strong enough and arranged on one surface so that the package can be plugged into a test or mounting socket and removed for replacement as desired without destruction.

Point-to-point wiring An interconnecting technique wherein the connections between components are made by wires routed between connecting points.

Polycrystalline A material is polycrystalline in nature if it is made of many small crystals. Alumina ceramics are polycrystalline, whereas glass substrates are not.

Polynary A material system with many basic compounds as ingredients, as thick-film resistor compositions. Binary indicates two compounds; ternary, three; etc.

Porosity The ratio of solid matter to voids in a material.

Positive image The true picture of a circuit pattern as opposed to the negative image or reversed image.

Positive temperature coefficient The changing of a value in the positive direction with increasing temperature.

Post See Terminal.

Post curing Heat aging of a film circuit after firing to stabilize the resistor values through stress relieving.

Post firing Refiring a film circuit after having gone through the firing cycle. Sometimes used to change the values of the already fired resistors.

Post stress electrical The application of an electrical load to a film circuit to stress the resistors and evaluate the resulting change in values.

Potting Encapsulating of a circuit in plastic.

Power density The amount of power dissipated from a film resistor through the substrate measured in $W/in.^2$.

Power dissipation The dispersion of the heat generated from a film circuit when a current flows through it.

Power factor The ratio of the actual power of an alternating or pulsating current as measured by a wattmeter to the apparent power as measured by an ammeter and voltmeter.

Prefired Conductors fired in advance of the screening of resistors on a substrate.

Preform To aid in soldering or adhesion, small circles or squares of the solder or epoxy are punched out of thin sheets. These preforms are placed on the spot to be soldered or bonded, prior to the placing of the object to be attached.

Preoxidized Resistor metal particles that have been oxidized to achieve the desired resistivity prior to formulation into a resistor ink.

Preseal visual The process of visual inspection of a completed hybrid circuit assembly for defects prior to sealing the package.

Pressed alumina Aluminum oxide ceramic formed by applying pressure to the ceramic powder and a binder prior to firing in a kiln.

Print and fire A term sometimes used to indicate steps in the thick-film process wherein the ink is printed on a substrate and is fired.

Print laydown Screening of the film circuit pattern onto a substrate.

Printing Same as Print laydown.

Printing parameters The conditions that affect the screening operation such as off-contact spacing, speed and pressure of squeegee, etc.

Probe A pointed conductor used in making electrical contact to a circuit pad for testing.

Procuring activity The organizational element (equipment manufacturer, government, contractor, subcontractor, or other responsible organization) which contracts for articles, supplies, or services and has the authority to grant waivers, deviations, or exceptions to the procurement documents.

Production lot Hybrid microcircuits manufactured on the same production line(s) by means of the same production techniques, materials, controls, and design. The production lot is usually date coded to permit control and traceability required for maintenance of reliability programs.

Profile (firing) A graph of time vs. temperature, or of position in a continuous thick-film furnace vs. temperature.

Property The physical, chemical, or electrical characteristic of a given material.

Pull strength The values of the pressure achieved in a test where a pulling stress is applied to determine breaking strength of a lead or bond.

Pull test A test for bond strength of a lead, interconnecting wire, or a conductor.

Pulse soldering Soldering a connection by melting the solder in the joint area by pulsing current through a high-resistance point applied to the joint area and the solder.

Purge To evacuate an area or volume space of all unwanted gasses, moisture, or contaminants prior to backfilling with an inert gas.

Purple plague One of several gold-aluminum compounds formed when bonding gold to aluminum and activated by re-exposure to moisture and high temperature ($>340°C$). Purple plague is purplish in color and is very brittle, potentially leading to time-based failure of the bonds. Its growth is highly enhanced by the presence of silicon to form ternary compounds.

Push-off strength The amount of force required to dislodge a chip device from its mounting pad by application of the force to one side of the device, parallel to the mounting surface.

Pyrolyzed (burned) A material that has gained its final form by the action of heat is said to be pyrolyzed.

Q The inverse ratio of the frequency band between half-power points (bandwidth) to the resonant frequency of the oscillating system. Refers to the electromechanical system of an ultrasonic bonder, or sensitivity of the mechanical resonance to changes in driving frequency.

Radiographs Photographs made of the interior of a sealed package by use of X-rays to expose the film.

Random failures Circuit failures which occur randomly with the overall failure rate for the sample population being nearly constant.

RC network A network composed only of resistors and capacitors.

Reactive metal Metals that readily form compounds.

Real estate The surface area of an integrated circuit or of a substrate. The surface area required for a component or element.

Rebond A second bonding attempt after a bond has been removed or failed to bond on the first attempt.

Rebonding-over bond A second bond made on top of a removed or damaged bond or a second bond made immediately adjacent to the first bond.

Reducing atmosphere An atmosphere containing a gas such as hydrogen that will reduce the oxidation state of the subject compound.

Reduction dimension A dimension specified on enlarged scale matrices, between a pair of marks which are positioned in very accurate alignment with the horizontal center locations of two manufacturing holes, and their locationally coincident targets. This dimension is used to indicate and verify the exact horizontal distance required between the two targets when the matrices are photographically reduced to full size.

Refiring Recycling a thick film resistor through the firing cycle to change the resistor value.

Glossary of Terms 195

Reflow soldering A method of soldering involving application of solder prior to the actual joining. To solder, the parts are joined and heated, causing the solder to remelt, or reflow.
Refractory metal Metals having a very high melting point, such as molybdenum.
Registration The alignment of a circuit pattern on a substrate.
Registration marks The marks used for aligning successive processing masks.
Reinforced plastic Plastics having reinforcing materials such as fiberglass embedded or laminated in the cured plastic.
Resist A protective coating that will keep another material from attaching itself or coating something, as in solder resist, plating resist, or photoresist.
Resistance weld The joining of two conductors by heat and pressure with the heat generated by passing a high current through the two mechanically joined materials.
Resistivity (ρ) A proportionality factor characteristic of different substances equal to the resistance that a centimeter cube of the substance offers to the passage of electricity, the current being perpendicular to two parallel faces. It is defined by the expression $R = \rho L/A$, where R is the resistance of a uniform conductor, L its length, A its cross-sectional area, and ρ its resistivity. Resistivity is usually expressed in ohm-centimeters.
Resistor drift The change in resistance of a resistor through aging, and usually rated as percent change per 1000 hr.
Resistor geometry The film resistor outline.
Resistor overlap The contact area between a film resistor and a film conductor.
Resistor paste calibration The characterizing of a resistor paste for Ω/square value, TCR and other specified parameters by screening and firing a test pattern using the paste and recording the results.
Resistor termination See Resistor overlap.
Resolution The degree of fineness or detail of a screen printed pattern (*see* Line definition).
Rework An operation performed on a nonconforming part or assembly that restores all nonconforming characteristics to the requirements in the contract, specifications, drawing or other approved product description.
Rheology The science dealing with deformation and flow of matter.
Ribbon interconnect A flat narrow ribbon of metal such as nickel, aluminum, or gold used to interconnect circuit elements or to connect the element to the output pins.
Ribbon wire Metal in the form of a very flexible flat thread or slender rod or bar tending to have a rectangular cross section as opposed to a round cross-section.

Rigid coating A conformal coating of thermosetting plastic that has no fillers or plasticizers to keep the coating pliable.

Risers The conductive paths that run vertically from one level of conductors to another in a multilayer substrate or screen printed film circuit.

Rosin flux A flux having a rosin base which becomes inactive after being subjected to the soldering temperature.

Rosin solder connection (rosin joint) A soldered joint in which one of the parts is surrounded by an almost invisible film of insulating rosin, making the joint intermittently or continuously open even though the joint looks good.

Sagging or wire sag The failure of bonding wire to form the loop defined by the path of the bonding tool between bonds.

Sapphire A single crystal Al_2O_3 substrate material used in integrated circuits.

Scaling Peeling of a film conductor or film resistor from a substrate, indicating poor adhesion.

Scallop marks A screening defect which is characterized by a print having jagged edges. This condition is a result of incorrect dynamic printing pressure or insufficient emulsion thickness.

Scavenging Same as Leaching.

Schematic Diagram of a functional electronic circuit composed of symbols of all active and passive elements and their interconnecting matrix that forms the circuit.

Scored substrate A substrate that has been scribed with a thin cut at the breaklines.

Screen A network of metal or fabric strands, mounted snuggly on a frame, and upon which the film circuit patterns and configurations are superimposed by photographic means.

Screen deposition The laydown of a circuit pattern on a substrate using the silk screening technique.

Screen frame A metal, wood, or plastic frame that holds the silk or stainless steel screen tautly in place.

Screening The process whereby the desired film circuit patterns and configurations are transferred to the surface of the substrate during manufacture, by forcing a material through the open areas of the screen using the wiping action of a soft squeegee.

Scribe coat A two-layer system, sometimes used for the preparation of thick-film circuit artwork, consisting of a translucent, dimensionally stable, polyester base layer, covered with a soft, opaque, strippable layer, and normally processed on a coordinatograph machine (*see* Cut and strip).

Scrubbing action Rubbing of a chip device around on a bonding pad during the bonding operation to break up the oxide layer and improve wetability of the eutectic alloy used in forming the bond.

Search height The height of the bonding tool above the bonding area at which final adjustments in the location of the bonding area under the tool are made prior to lowering the tool for bonding.

Second bond The second bond of a bond pair made to form a conductive connection.

Second radius The radius of the back edge of the bonding tool foot.

Second search That period of machine cycle at which final adjustments in the location of the second bonding area (*see* Second bond) under the tool are made prior to lowering the tool for making the second wire bond.

Selective etch Restricting the etching action on a pattern by the use of selective chemical which attack only one of the exposed materials.

Self-heating Generation of heat with a body by chemical action. Epoxy materials self-heat in curing due to exothermic reaction.

Self-passivating glaze The glassy material in a thick-film resistor that comes to the surface and seals the surface against moisture.

Semiconductor carrier A permanent protective structure which provides for mounting and for electrical continuity in application of a semiconductor chip to a major substrate.

Semiconductors Solid materials such as silicon that have a resistivity midway between that of a conductor and a resistor. These materials are used as substrates for semiconductor devices such as transistors, diodes, and integrated circuits.

Serpentine cut A trim cut in a film resistor that follows a serpentine or wiggly pattern to effectively increase the resistor length and increase resistance.

Shear rate The relative rate of flow or movement (of viscous fluids).

Shear strength The limiting stress of a material determined by measuring a strain resulting from applied forces that cause or tend to cause contiguous parts of a body to slide relative to each other in a direction parallel to their plane of contact; the value of the force achieved when shearing stress is applied to the bond (normally parallel to the substrate) to determine the breaking load.

Sheet resistance The electrical resistance of a thin sheet of a material with uniform thickness as measured across opposite sides of a unit square pattern. Expressed in Ω/square.

Shelf life The maximum length of time, usually measured in months, between the date of shipment of a material to a customer and the date by which it should be used for best results.

Short-term overload A circuit that has been overloaded with current or voltage for a period too short to cause breakdown of the insulation.

Silicon monoxide A passivating or insulating material that is vapor deposited on selected areas of a thin film circuit.

Silk screen A screen of a closely woven silk mesh stretched over a frame and used to hold an emulsion outlining a circuit pattern and used in screen printing of film circuits. Used generically to describe any screen (stainless steel or nylon) used for screen printing.

Sinking Shorting of one conductor to another on multilayer screen printed circuits because of a downward movement of the top conductor through the molten crossover glass.

Sintering Heating a metal powder under pressure and causing the particles to bond together in a mass. Alternately, subjecting a ceramic-powder mix to a firing cycle whereby the mix is less than completely fused and shrinks.

Skin effect The increase in resistance of a conductor at microwave frequencies because of the tendency for current to concentrate at the conductor surface.

Slice A thin corss section of a crystal such as silicon that is used for semiconductor substrates.

Slump A spreading of printed thick-film composition after screen printing but before drying. Too much slumping results in loss of definition.

Slurry A thick mixture of liquid and solids the solids being in suspension in the liquid.

Smeared bond A bond impression that has been distorted or enlarged by excess lateral movement of the bonding tool, or by the movement of the device-holding fixture.

Snapback The return of a screen to normal after being deflected by the squeegee moving across the screen and substrate.

Snap-off distance The screen printer distance setting between the bottom of the screen and top of the substrate (*see* Breakaway).

Snapstrate A scribed substrate that can be processed by gang deposition in multiples of circuits and snapped apart afterward.

Soak time The length of time a ceramic material (such as a substrate or thick-film composition) is held at the peak temperature of the firing cycle.

Soft solder A low-melting solder, generally a lead-tin alloy, with a melting point below 800°F (425 °C).

Solder acceptance Same as Wetability or Solderability.

Solder bumps The round solder balls bonded to a transistor contact area and used to make connection to a conductor by face-down bonding techniques.

Solder cream Solder paste.

Solder dam A dielectric composition screened across a conductor to limit molten solder from spreading further onto solderable conductors.

Solder glasses Glasses used in package sealing that have a low melting point and tend to wet metal and ceramic surfaces.

Solder immersion A test that immerses the electronic package leads into a solder bath to check resistance to soldering temperatures.
Solder resist A material used to localize and control the size of soldering areas, usually around component mounting holes. The solderable areas are defined by the solder resist matrix.
Solderability The ability of a conductor to be wetted by solder and to form a strong bond with the solder.
Soldering The process of joining metals by fusion and solidification of an adherent alloy having a melting point below about 800°F.
Solid metal mask A thin sheet of metal with an etched pattern used in contact printing of film circuits.
Solid phase bond The formation of a bond between two parts in the absence of any liquid phase at any time prior to or during the joining process.
Solid state Pertaining to circuits and components using semiconductors as substrates.
Solid tantalum chip A chip or leadless capacitor whose dielectric (Ta_2O_5) is formed with a solid electrolyte instead of a liquid electrolyte.
Solidus The locus of points in a phase diagram representing the temperature, under equilibrium conditions, at which each composition in the system begins to melt during heating, or complete freezing during cooling.
Solubility The ability of a substance to dissolve into a solvent.
Solvent A material that has the ability to dissolve other materials.
Solvent resistant A material that is unaffected by solvents and does not degrade when cleaned in solvents.
Spacings The distance between adjacent conductor edges.
Specific gravity The ratio of the weight of a given volume of a substance to the weight of an equal volume of water at a temperature of 4°C.
Specific heat The quantity of heat required to raise the temperature of 1 g of a substance 1°C.
Specimen A sample of material, device or circuit representing the production lot removed for test.
Spikes See Curls.
Spinel A single-crystal magnesium aluminum oxide substrate used in integrated circuits.
Split-tip electrode Same as Parallel-gap electrode.
Sputtering The removal of atoms from a source by energetic ion bombardment, the ions supplied by a plasma. Process is used to deposit films for various thin-film applications.
Squashout The deformed area of a lead which extends beyond the dimensions of the lead prior to bonding.
Squeegee The part of a screen printer that pushes the composition across the screen and through the mesh onto the substrate.

Stainless steel screen A stainless steel mesh screen stretched across a frame and used to support a circuit pattern defined by an emulsion bonded to the screen.

Stair-step print A print which retains the pattern of the screen mesh at the line edges. This is a result of inadequate dynamic printing pressure exerted on the paste or insufficient emulsion thickness coating the screen.

Standard deviation A statistical term that helps describe the likely value of parts in a lot or batch of components in comparison with the lot's average value. Practically all of a lot will fall within ±3 standard deviations of the average value if it is a normal distribution.

Standoff A connecting post of metal bonded to a conductor and raised above the surface of the film circuit.

Steatite A ceramic consisting chiefly of a silicate of magnesium used as an insulator or circuit substrate.

Stencil A thin sheet material with a circuit pattern cut into the material. A metal mask is a stencil.

Step-and-repeat A process wherein the conductor or resistor pattern is repeated many times in evenly spaced rows onto a single film or substrate.

Stitch bond A bond made with a capillary-type bonding tool when the wire is not formed into a ball prior to bonding.

Stratification The separation of nonvolatile components of a thick film into horizontal layers during firing, due to large differences in density of the component. It is more likely to occur with a glass containing conductor paste, and under prolonged, or repeated firing.

Stray capacitance Capacitance developed from adjacent conductors separated by an air dielectric or dielectric material.

Stress relieve A process of reheating a film resistor to make it stress free.

Stress-free The annealed or stress relieved material where the molecules are no longer in tension.

Stripline A microwave conductor on a substrate.

Stylus A sharp-pointed probe used in making an electrical contact on the pad of a leadless device or a film circuit.

Subcarrier substrate A small substrate of a film circuit which is mounted in turn to a larger substrate.

Substrate (of a microcircuit or integrated circuit) The supporting material upon which the elements of a hybrid microcircuit are deposited or attached or within which the elements of an integrated circuit are fabricated.

Subsystem A smaller part of an electronic system which performs a part of the system function but can be removed intact and tested separately.

Surface conductance Conductance of electrons along the outer surface of a conductor.
Surface diffusion The high-temperature injection of atoms into the surface layer of a semiconductor material to form the junctions. Usually, a gaseous diffusion process.
Surface finish The peaks and valleys in the surface of a substrate rated in μin./in. deviation.
Surface nucleation The change in phase or state of the surface on a substrate.
Surface resistivity The resistance to a current flow along the surface of an insulator material.
Surface tension An effect of the forces of attraction existing between the molecules of a liquid. It exists only on the boundary surface.
Surface texture The smoothness or lack of it on the surface of a substrate.
Surfactant A contraction for the term "surface-active agent."
Swimming Lateral shifting of a thick-film conductor pattern on molten glass crossover patterns.
Tacky state A material is in a tacky state when it exhibits an adhesive bond to another surface.
Tail (of the bond) The free end of wire extending beyond the bond impression of a wire bond from the heel.
Tail pull The act of removing the excess wire left when a wedge or ultrasonic bond is made.
Tantalum capacitor Capacitors that utilize a thin tantalum oxide layer as the dielectric material.
Tape alumina Alumina (substrates) made by tape casting of slurry into strips of green alumina of a predetermined thickness. This is followed by stamping, cutting in the green state, then firing.
Tarnish Chemical accretions on the surface of metals, such as sulfides and oxides. Solder fluxes have to remove tarnish in order to allow wetting.
Tear strength Measurement of the amount of force needed to tear a solid material that has been nicked on one edge and then subjected to a pulling stress. Measured in lb/in.
Temperature aging Aging or stressing a film circuit in an elevated temperature over a period of time.
Temperature coefficient of capacitance (TCC) The amount of capacitance change of a capacitor with temperature, commonly expressed as the average change over a certain temperature range in ppm/°C.
Temperature coefficient of resistance (TCR) The amount of resistance change of a resistor (or resistor material) with temperature, commonly expressed as the average change over a certain temperature range in ppm/°C.

Temperature cycling An environmental test where the film circuit is subjected to several temperature changes from a low temperature to a high temperature over a period of time.

Temperature excursion The extreme temperature differences seen by a film circuit under operating conditions.

Temperature tracking The ability of a component to retrace its electrical readings going up and down the temperature scale.

Tensile strength The pulling stress which has to be applied to a material to break it, usually measured in psi.

Terminal A metal lead used to provide electrical access to the inside of the device package.

Test pattern A circuit or group of substrate elements processed on or within a substrate to act as a test site or sites for element evaluation or monitoring of fabrication processes.

Thermal conductivity The rate with which a material is capable of transferring a given amount of heat through itself.

Thermal design The schematic heat flow path for power dissipation from within a film circuit to a heat sink.

Thermal drift The deift of circuit elements from nominal value due to changes in temperature.

Thermal drop The difference in temperature across a boundary or across a material.

Thermal gradient The plot of temperature variances across the surface or the bulk thickness of a material being heated.

Thermal mismatch Differences of thermal coefficients of expansion of materials which are bonded together.

Thermal noise Noise that is generated by the random thermal motion of charged particles in an electronic device.

Thermal runaway A condition wherein the heat generated by a device causes an increase in heat generated. This spiraling rise in dissipation usually continues until a temperature is reached that results in destruction of the device.

Thermal shift The permanent shift in the nominal value of a circuit element due to heating effect.

Thermal shock A condition whereby devices are subjected alternately to extreme heat and extreme cold. Used to screen out processing defects.

Thermocompression bonding A process involving the use of pressure and temperature to join two materials by interdiffusion across the boundary.

Thermoplastic A substance that becomes plastic (malleable) on being heated; a plastic material that can be repeatedly melted or softened by heat without change of properties.

Thermoswaging Heating a pin that is inserted in a hole and upsetting the hot metal so that it swells and fills the hole, thereby forming a tight bond with the base material.

Thick film A film deposited by screen printing processes and fired at high temperature to fuse into its final form. The basic processes of thick-film technology are screen printing and firing.

Thick-film circuit A microcircuit in which passive components of a ceramic-metal composition are formed on a suitable substrate by screening and firing.

Thick-film hybrid circuit A hybrid microcircuit that has add-on components, usually chip devices added to a thick-film network to perform an electronic function.

Thick-film network A network of thick-film resistors and/or capacitors interconnected with thick-film conductors on a ceramic substrate, formed by screening and firing.

Thick-film technology The technology whereby electrical networks or elements are formed by applying a liquid, solid, or paste coating through a screen or mask in a selective pattern onto a supporting material (substrate) and fired. Films so formed are usually 5 µm or greater in thickness.

Thin film A thin film (usually less than 1000 µm thickness) is one that is deposited onto a substrate by an accretion process such as vacuum evaporation, sputtering, or pyrolytic decomposition.

Thin-film hybrid circuit A hybrid microcircuit that has add-on components, usually chip devices added to a thin-film network to perform an electronic function.

Thin-film integrated circuit A functioning circuit made entirely of thin-film components. Also used to mean thin-film hybrid circuit.

Thin-film network A resistor and/or capacitor and conductor network formed on a single substrate by vacuum evaporation and sputtering techniques.

Thin-film technology The technology whereby electronic networks or elements are formed by vacuum evaporation or sputtering films onto a supporting material (substrate). Films so formed are less than 5 and usually of the order 0.3 to 1.0 µm in thickness.

Thixotropic A fluid that gets less viscous as it is stirred (or moved) is pseudoplastic. The term is sometimes applied to these fluids (e.g., most thick-film inks).

Throwaway module A functional circuit in a modular form factor that is considered expendable and will not be repaired because of its low cost.

Tinned Literally, coated with tin, but commonly used to indicate coated with solder.

Tinning To coat metallic surfaces with a thin layer of solder.

Tip That portion of the bonding tool which deforms the wire to cause the bond impression.

TO package Abbreviation for transistor outline, established as an industry standard by JEDEC of the EIA.

Toe See Tail (of the bond).

Top hat resistors Film resistors having a projection out one side allowing a notch to be cut into the center of the projection to form a serpentine resistor and thereby increase the resistivity.

Topography The surface condition of a film—bumps, craters, etc.

Topology The surface layout design study and characterization of a microcircuit. It has application chiefly in the preparation of the artwork for the layout masks used in fabrication.

Toroids A helical winding on a ring-shaped core (doughnut-shaped coil). A popular form used for inductors and transformers in hybrid microcircuits because of their volumetric efficiency compared to other shapes.

Torque test A test for determining the amount of torque required to twist off a lead or terminal.

Tracking Two similar elements on the same circuit that change values with temperature in close harmony are said to track well. Tracking of different resistors is measured in ppm/°C (difference). Tracking is also used in reference to temperature hysteresis performance and potentiometer repeatability.

Transducer A device actuated by one transmission system and supplying related energy to another transmission system.

Transfer molded Molding circuit modules by transferring molten plastic into a cavity holding the circuit by using a press.

Trapezoidal distortion Distortion that can occur during photoreduction. As a result, a square shape in the original master will be transformed into a trapezoid at the reduced positive.

Trim notch The notch made in a resistor by trimming to obtain the design value (*see* Kerf).

Trimming Notching a resistor by abrasive or laser means to raise the nominal resistance value.

Ultrasonic bonding A process involving the use of ultrasonic energy and pressure to join two materials.

Ultrasonic cleaning A method of cleaning that uses cavitation in fluids caused by applying ultrasonic vibrations to the fluid.

Ultrasonic power supply An electronic high-frequency generator that provides ultrasonic power to a transducer.

Uncased device A chip device.

Underdeformed Insufficient deformation of the wire by the bonding tool occurring during the bonding operation.

Underglaze A glass or ceramic glaze applied to a substrate prior to the screening and firing of a resistor.

Universal lead frame Lead frame in which flag is not supported by any lead.

Vacuum deposition Deposition of a metal film onto a substrate in a vacuum by metal evaporation techniques.

Vacuum pickup   A handling instrument with a small vacuum cup on one end used to pick up chip devices.
Vapor deposition (vapor evaporation)   Same as Vacuum deposition
Vapor phase   The state of a compound when it is in the form of a vapor.
Varnish   A protective coating for a circuit to protect the elements from environmental damage.
Vehicle   A thick-film term that refers to the organic system in the paste.
Via   An opening in the dielectric layer through which a riser passes.
Vintage control number   An alphanumeric code relating to changes made on manufactured products. The vintage control code generally follows the item part number.
Viscosimeter (viscometer)   A device that measures viscosity. Viscometers for thick-film compositions must be capable of measuring viscosity under conditions of varying shear rates.
Viscosity   A term used to describe the fluidity of material, or the rate of flow versus pressure. The unit of viscosity measurement is poise, more commonly centipoise. Viscosity varies inversely with temperature.
Viscosity coefficient   The coefficient of viscosity is the value of the tangential force per unit area which is necessary to maintain unit relative velocity between two parallel planes unit distance apart.
Vitreous   A term used in ceramic technology indicating fired characteristics approaching being glassy, but not necessarily totally glassy.
Vitreous binder   A glassy material used in a compound to bind other particles together. This takes place after melting the glass and cooling.
Vitrification   The reduction of porosity in a ceramic product through the formation of a glassy bond.
Voltage gradient   The voltage drop (or change) per unit length along a resistor or other conductance path.
Voltage rating   The maximum voltage which an electronic circuit can sustain to ensure long life and reliable operation.
Volume resistivity   A 1-cm cube of material will have resistance equal to the material's resistivity. The qualification "volume" adds nothing, but is sometimes used so that "resistivity" and "resistance" will not be confused.
Wafer   A slice of semiconductor crystal ingot used as a substrate for transistors, diodes, and monolithic integrated circuits.
Warp and woof   Threads in a woven screen which cross each other at right angles.
Warpage   The distortion of a substrate from a flat plane.
Waves   Any disturbance that advances through a medium with a speed determined by properties of the medium.

Waviness One or a series of elevations or depressions or both, which are readily noticeable and which include defects such as buckles, ridges, etc.

Wedge bond A bond made with a wedge tool. The term is usually used to differentiate thermocompression wedge bonds from other thermocompression bonds. (Almost all ultrasonic bonds are wedge bonds.)

Wetting The spreading of molten solder on a metallic surface, with proper application of heat and flux.

Wicking The flow of solder along the strands and under the insulation of stranded lead wires.

Wire bond Includes all the constituent components of a wire electrical connection such as between the terminal and the semiconductor die. These components are the wire, metal bonding surfaces, the adjacent underlying insulating layer (if present), and substrate.

Wire bonding The method used to attach very fine wire to semiconductor components to interconnect these components with each other or with package leads.

Wire clamp A device designed to hold the wire during the cutoff operation.

Wire spool The wire magazine.

Wobble bond A thermocompression, multicontact bond accomplished by rocking (or wobbling) a bonding tool on the beams of a beam lead device.

Worst-case analysis The analysis of a circuit function under tolerance extremes of temperature, humidity, etc., to determine the worst possible effect on the output parameters.

Woven screen A screen mesh used for screen printing usually of nylon or stainless steel or possibly silk.

Yield The ratio of usable components at the end of a manufacturing process to the number of components initially submitted for processing. Can be applied to any input-output stage in processing, and so must be carefully defined and understood.

NEW TERMS

Bond separation The distance between the attachment points of the first and second bonds of a wire bond.

BTAB The acronym for tape automated bonding when the raised bump for each bond site is prepared on the tape material as opposed to the bump being on the component.

Chip carrier A special type of enclosure or package used to house a semiconductor device or a hybrid microcircuit which has metallized electrical terminations around its perimeter rather than an extended lead frame or plug-in pins.

Glossary of Terms

Devitrification The action of process of devitrifying or state of being devitrified—the conversion of a glassy matter into crystalline.

Devitrify To deprive of glassy luster and transparency—to change from a vitreous to a crystalline condition.

Gang bonding The act of bonding a plurality of mechanical and/or electrical connections through a single act or stroke of a bonding tool.

Glass transition temperature (Tg) In polymer chemistry, the temperature below which the thermal expansion coefficient becomes nearly constant or a simple function of temperature.

Gram-force A unit of force (nominally 9.8 mN) required to support a mass of one gram (1 gravity unit of acceleration times one gram of mass equals 1 gram-force). Colloquially, the term gram is used for the unit.

ILB The acronym for inner lead bond as referring to the connection between the bonding tape terminals and the electronic component, usually the first gang bond to occur.

Junction (1) In solid state materials, a region of transition between p- and n- type semiconductor material as in a transistor or diode. (2) A contact between two dissimilar metals or materials (e.g., in a thermocouple or rectifier). (3) A connection between two or more conductors or two or more sections of a transmission line.

MTBF The acronym for mean time between failures applied statistically to reliability studies.

OLB The acronym for the second gang bonding process, outer lead bond, which attaches the tape bonded component to the substrate—the next higher assembly level.

Repair An operation performed on a nonconforming part or assembly to make it functionally usable but which does not completely eliminate the nonconformance.

Scratch In optical observations, a surface mark with a large length-to-width ratio.

TAB The acronym for tape automated bonding.

Tape bonding The utilization of a metal or plastic tape material as a support and carrier of a microelectronic component in a gang bonding process.

Underbonding In wire bonding, insufficiently deforming the wire, with the bonding tool, during the bonding process.

Vitreous Having the nature of glass.

Vitrification The progressive reduction in porosity of a ceramic material as a result of heat treatment or some other process.

Void In visual inspection of solid materials, a space not filled with the specific solid material such as a gap or opening which is an unintentional defect in the material.

Wedge tool A bonding tool in the general form of a wedge with or without a wire-guide hole to position the wire under the bonding face of the tool, as opposed to a capillary-type tool.

INDEX

A

Acceleration factor, 154-155
Accelerated life tests, 149
Alumina, 10, 13
Applications, 143
Assembly
 rates, 121
 yields, 122

B

Beam lead
 attachment, 110
 devices, 98
Beryllia, 13
Binder, 22, 23, 27, 30
Bonding
 automated wire, 114
 inner lead, 112
 outer lead, 112
 thermocompression, 107-109
 ultrasonic, 109-110
 wobble, 111

C

Camber, 10
Capacitor
 interdigital, 59, 76
 parallel plate, 14, 59, 77
 substrate, 58, 76

[Capacitor]
 thick film characteristics, 60
 thin film characteristics, 78
 trimming, 61, 78
Center line average (CLA), 17
Ceramic covers, 115
Chip, 6
Chip carriers, 101
Circuit analysis, 144-146
Conductors
 copper, 25
 gold, 24
 multilayer, 26
 nickel, 25
 palladium gold, 24
 palladium silver, 24
 platinum gold, 24
 silver, 24
 thick-film characteristics, 48
 thick-film crossovers, 56-58
 thick-film design, 48-58
 thin-film design, 74-75
Conformal coatings, 116
Constant failure rate, 153, 155-157

D

Design cycle, 140, 145
Development
 cost, 3, 6, 141
 time, 6

Die (dice), 6
Dielectric
  constant, 9, 10, 13
  strength, 9, 14
  types of, 28
Dissipation
  factor, 9, 15

E

Economic aspects, 3
Electrical properties, 3
Epoxy die attach, 107
Etched circuit board, 4
Eutectic die attach, 106

F

Failure
  analysis, 162
  distribution, 153
  infant, 153
  mechanisms, 161-162
  multiple element systems, 158
  rate, 153
  wear-out, 153
Firing
  substrate, 20
  thick-film pastes, 33-36
Flip chip devices, 98
Fluidized-bed, 117

G

Glazing, 20
Green tape, 20

H

Hermetic packages, 115
High-temperature storage, 148, 164
Hybrid microcircuit, 1

I

Inductors, 97
Integrated circuits, (IC), 6
Isotherms, 129

L

Laser
  scribing, 21
  trimming, 81
  trim system, 83-84
Layout, 146-147
Leadless inverted device (LID), 98
Leak rate, 115
Life tests, 148
Loss angle, 15

M

Mean time between failures, (MTBF), 157
Mean time to failure (MTTF), 157-158
Micropackages, 100
Microwave circuits, 1-2
Milestones
  design review, 150
  engineering completion, 151
  feasibility, 150
  release to production, 151
Monolithic circuits, 6

N

Nickel chromium, 67

P

Partitioning, 138-139
Parylene, 117
Plating
  electroless, 65
  electroplating, 65
Polyimide, 71
Pressing
  dry, 19
  powder, 19
Printed circuit board, 4
Probe system
  four-point, 85
  two-point, 85

Index

R

Reliability definitions, 153-155
Reliability goals, 158-159
Reliability testing, 159-161
Resistivity
  sheet, 25, 40
  volume, 9, 17
Resistors
  thick film, 29-31
  thick-film characteristics, 46
  thick-film design, 40-46
  thin-film characteristics, 73
  thin-film design, 72-74
  trim geometries, 43-44, 74

S

Screening, 31-33, 121
Sheet casting, 20
Snapstrates, 79
Solder
  dam, 104
  fluxes, 103
  materials, 103
  processes, 104
Specifications
  hybrid, 119
  material, 118
  process, 118
Sputtering
  cathode, 64
  DC, 64
  RF, 65
Statistical factor (MTTF), 160-161
Substrate
  alumina, 10
  beryllia, 10
  definition, 9
  engineering concerns, 9
  fabrication, 19
  fosterite, 10
  glass, 13
  manufacturing concerns, 9
  materials, 10

[Substrate]
  porcelain steel, 11
  quartz, 13
  sapphire, 13
  stability, 17
  steatite, 10
  surface, 17
  tolerances, 10, 18
Surface
  finish, 10
  flatness, 18
  substrate, 17

T

Tantalum nitride, 67
Tape carrier, 112
Temperature cycling, 148, 164
Testing, 149-150
Thermal
  calculations, 131-133
  coefficient of expansion, 9
  conductivity, 9, 13, 126, 133
  conductivity of materials, 134
  dissipation, 6
  flow, 127
  resistance, 127-128
  shock, 148, 164
Thermal Design
  conduction equation, 126
  guidelines, 136-137
  rules of thumb, 125
Thick film
  conductors, 22-27
  definition of, 8
  dielectrics, 27-29
  dimensions, 7
  inks, 22
  pastes, 22
  processing, 31-36
  resistors, 29-31
Thin film
  conductors, 66
  definition of, 8
  dielectrics, 67
  dimensions, 7

[Thin film]
  environmental protection of, 71
  processing, 67-71
  resistors, 67
Tolerances
  substrate, 10
Top-hat resistors
  trim information, 88-94
Trimming
  air abrasive, 81
  functional, 84
  laser, 81
  predictive, 84
Trim ratio, 43

V

Vacuum evaporation, 62-64
Vapor-phase deposition, 65
Vehicle, 22, 23, 28, 30
Visual inspection, 120

W

Wobble bonding, 111

Y

Yield, 6